Geology of Lode Gold Districts in the Klamath Mountains, California and Oregon

By PRESTON E. HOTZ

GEOLOGICAL SURVEY BULLETIN 1290

UNITED STATES GOVERNMENT PRINTING OFFICE, WASHINGTON : 1971

UNITED STATES DEPARTMENT OF THE INTERIOR

GEOLOGICAL SURVEY

William T. Pecora, *Director*

Library of Congress catalog-card No. 76–608514

For sale by the Superintendent of Documents, U. S. Government Printing Office
Washington, D. C. 20402

CONTENTS

ILLUSTRATIONS

TABLES

GEOLOGY OF LODE GOLD DISTRICTS IN THE KLAMATH MOUNTAINS, CALIFORNIA AND OREGON

By Preston E. Hotz

ABSTRACT

The Klamath Mountains geologic province, in northwestern California and southwestern Oregon, is divisible into four north-trending concentric arcuate lithologic belts which are convex to the west. These belts are designated, from east to west, (1) the eastern Klamath belt, (2) the central metamorphic belt, (3) the western Paleozoic and Triassic belt, and (4) the western Jurassic belt. Rocks of the eastern Klamath belt range in age from Ordovician (?) to Jurassic and include typical eugeosynclinal clastic sediments and volcanic rocks. They have an aggregate thickness of approximately 40,000–50,000 feet.

The Salmon Hornblende Schist and the Abrams Mica Schist, whose initial metamorphic age is Devonian as determined by rubidium-strontium techniques, make up the central metamorphic belt.

The western Paleozoic and Triassic belt is the most extensive unit in the Klamath Mountains and is a typical eugeosynclinal assemblage of fine-grained clastic sediments, chert, mafic volcanic rocks, and lenticular marble, all regionally metamorphosed under conditions of the lower greenschist facies. Meager fossil data indicate that the age of the rocks ranges from Early Permian to Late Triassic. Metamorphic equivalents of the western Paleozoic and Triassic belt that occur along its east border and as fensters in the central metamorphic belt constitute the Stuart Fork Formation of Davis and Lipman (1962), an assemblage of highly deformed siliceous schists, greenstones, marble, and blueschist facies rocks. Large areas of amphibolite and siliceous metasedimentary rocks of the almandine-amphibolite facies within the western Paleozoic and Triassic belt are tentatively believed to be higher grade equivalents of the greenstones and siliceous sediments. A subcircular window exposes graphitic micaceous schist and actinolite schist, called the schists of Condrey Mountain. These rocks underlie the higher grade metamorphic units. They were metamorphosed in Late Jurassic time, but their parental equivalents are unknown.

The western Jurassic belt is composed of slate and graywacke of the Late Jurassic Galice Formation and volcanic rocks that range in composition from basalt to rhyolite.

The four lithologic belts are bounded by thrust faults along which each belt has overridden its western neighbor. In each belt, the bedding and (or) metamorphic foliation are inclined to the east. The beds are complexly folded, and commonly the axial planes dip eastward.

Ultramafic bodies and granitic plutons occur in all the lithologic belts. Ultramafic rocks range from pyroxenite to dunite, but serpentinized peridotite is the most common type. The ultramafic bodies are commonly elongate and concordant with the structural grain of the province; outcrops range in area from a few acres to hundreds of square miles. Several long, continuous bodies of serpentinite occupy the thrusts bounding the major lithologic belts; and two large bodies, one in the southeastern part of the province and one in the western part, are interpreted as sheetlike plutons, one separating the eastern Klamath from the central metamorphic belt and the other separating the western Paleozoic and Triassic belt from the Jurassic belt. Many of the small irregularly shaped bodies are possibly remnants of formerly more continuous tabular bodies that were infolded with the rocks they intrude. Mafic rocks that range in composition from diabase to gabbro accompany the ultramafic rocks.

Granitic plutons range in size from small stocks to batholiths with outcrop areas of 100 square miles or more. The plutons also tend to be elongate north-south, parallel to the prevailing structural grain. Quartz diorite is the predominant rock type; diorite and granodiorite are common, whereas gabbro and quartz monzonite are rare. Some plutons, especially in the south-central part of the province, are trondhjemitic in part. Isotopic ages for most of the plutons are Middle and Late Jurassic and, with rare exception, define three episodes of intrusion: 165–167 m.y. (million years), 145–155 m.y., and 127–140 m.y. A small stock near Redding has a minimum age of 246 m.y. (Permian).

Marine sedimentary rocks of Cretaceous age overlap the older, subjacent rocks at the north, northeast, and southeast extents of the province. Tertiary units within the province include the Weaverville Formation in the southern Klamath Mountains and the Wimer Beds of Diller (1902) in the northwest. Volcanic rocks of Tertiary and Quaternary ages, belonging to the Cascade Range province, overlap the older rocks of the Klamath Mountains all along their eastern border. Quaternary alluvial deposits occupy the stream valleys in the province and have been source of much placer gold.

The first discovery of gold in the Klamath Mountains was a placer deposit in Shasta County, Calif., in 1848. In southwestern Oregon the earliest discoveries were in Jackson County in 1852. The first lode mining commenced in 1852 in Shasta County. More than 7 million ounces of gold, valued at more than $165 million, have been recovered from the placers and lode mines of the Klamath Mountains.

Most of the lode deposits of gold are simple quartz-filled fissure veins. The major mineral, quartz, is accompanied by subordinate amounts of carbonate. Metallic gangue minerals range in amount from less than 3 percent to as much as 5 percent and included pyrite and lesser amounts of arsenopyrite, galena, sphalerite, chalcopyrite, pyrrhotite, molybdenite, and, rarely, tellurides. Gold occurs free in quartz and in the sulfides, chiefly pyrite. Most of the mined veins contained an average of 0.1 to 0.4 ounce gold per ton. In the most successful mines, the gold content was approximately 0.4 to 0.75 ounce per ton.

Gold-sulfide replacement deposits are less common, but have been mined profitably in a few places as residually enriched gossans. The sulfides are mainly fine-grained pyrite with subordinate amounts of chalcopyrite, sphalerite, and galena. Grade of the unweathered protore is low, but the gold content of the gossan is several tenths of an ounce per ton.

An uncertain amount of the gold was mined from small, localized near-surface concentrations known as "pocket deposits." Many of these were very rich, but most were abandoned after the easily available gold was removed.

All rock types of the Klamath Mountains are hosts for the gold lodes. Metavolcanic rocks are the most common host, possibly because they are the most abundant rock in the Klamath Mountains. Some of the richest deposits were found in argillaceous sedimentary rock. In general, granitic rocks do not contain gold ore, although a few deposits near the borders of plutons are known. Rare gold-quartz veins crop out near and in serpentine.

Dike rocks are plentiful near many lode deposits; they may form one or both walls of a vein and, in places they seem to have been favored loci for gold-bearing veins. The most common dike rock near gold veins is the so-called birdseye porphyry, of diorite to quartz diorite composition. Basaltic and lamprophyric dikes have also been considered as indicators for auriferous veins. However, no genetic relationship between dike and gold-bearing veins has been proved, and the association may be only a structural relationship.

The distribution of some mineralized districts correlates with the known regional geology; others show no such relation. The most promising correlation is with thrust faults. The largest and most productive district, the French Gulch–Deadwood, is along an east-west fracture system near the crest of a broad arch in rocks of the eastern Klamath belt. Several deposits in this area occur in a thrust fault between metavolcanic rocks and slates of the eastern Klamath belt. Similarly, many deposits in the east-central part of the province occur in rocks of the western Paleozoic and Triassic belt near the thrust-fault contact with rocks of the central metamorphic belt. The deposits of the Greenback district in Oregon are near the thrust contact between rocks of the western Paleozoic and Triassic belt and the western Jurassic belt. Several deposits occur in the eastern Klamath belt adjacent to the north and south boundaries of the large Trinity ultramafic body which occupies the contact between the eastern Klamath belt and the central metamorphic belt.

In southern Oregon, however, lode deposits are scattered with no obvious pattern through a wide area of the western Paleozoic and Triassic belt; it appears that fracturing led to mineralization in many small veins instead of concentration in definite belts of lodes. There are few deposits in the western Jurassic belt, except in the Galice district where there was extensive faulting and emplacement of gabbroic and granitic rocks.

No deposits are known in the Dothan and Franciscan Formations west of the Klamath Mountains province: the small Mule Creek area in Oregon, west of the Klamath Mountains province, is in a small outcrop of volcanic rocks which are possibly correlative with volcanic rocks of the western Jurassic belt.

Although the province is widely intruded by granitic plutons, there is no regular, obvious relation between distribution of lode gold districts and exposures of granitic rocks.

Geologic evidence suggests that the gold lodes are genetically related to the granitic rocks of Late Jurassic age and that they are younger than Latest Jurassic or Early Cretaceous.

INTRODUCTION

This report summarizes an investigation of the geologic environments of the lode gold deposits in the Klamath Mountains, northern California and southwestern Oregon. Previously published literature contains many descriptions of individual mines and mining districts in this region, but few studies treat the broader relation of mines and districts to the regional geology. In recent years the geology of the Klamath Mountains province has become better known, and it is now possible to present a more complete description of the geologic environment in which the lode deposits of gold occur.

Descriptions of individual mines and districts have been published by the California Division of Mines and Geology and the Oregon Department of Geology and Mineral Industries. Geologic studies oriented toward gold mineralization in limited areas have been made by Albers (1965), Ferguson (1914), Hershey (1900), and MacDonald (1913) in California. Similar studies in Oregon were made by Diller (1914), Diller and Kay (1908; 1924), Shenon (1933a, b), Winchell (1914), and Youngberg (1947). The geology of the Klamath Mountains, known mainly from reconnaissance studies and local more detailed mapping, has been compiled and published at a scale of 1:250,000 in California (Strand, 1962, 1964) and at a scale of 1:500,000 in Oregon (Wells and Peck, 1961). The geologic maps (pls. 1, 2) incorporate many of these data.

KLAMATH MOUNTAINS PROVINCE

The Klamath Mountains geologic province occupies an elongate north-south area of approximately 12,000 square miles in northwestern California and southwestern Oregon (fig. 1). It is mainly underlain by Paleozoic and Mesozoic sedimentary, volcanic, and metamorphic rocks, in which there are numerous ultramafic bodies, and gabbroic, and granitic plutons. It is bordered on the east by Tertiary and Quaternary rocks of the Cascade province, on the southeast by Cretaceous sedimentary rocks of the Great Valley province of California, and on the west by the northern Coast Range province, which is underlain in California by sedimentary and volcanic rocks of the Late Jurassic (Tithonian) to Late Cretaceous Franciscan Formation and in Oregon by presumably partly equivalent rocks called the Dothan Formation.

The Klamath Mountains province is a highly dissected, rugged, mountainous terrain. Ridge crests commonly are 5,000–7,000 feet above sea level, and locally peaks attain altitudes of as much as 9,000 feet. Many of the river canyons have steep slopes, with a local relief of 3,000–4,000 feet. Vestiges of an old upland surface are recognizable, especially in the western part of the province where ridges have accordant summit levels and the dissected remains of broad valleys are visible. Diller (1902) was one of the first to recognize these

FIGURE 1.—Index map showing location of the Klamath
Mountains province.

physiographic features, and he named the old erosion surface the
Klamath peneplain.

The major streams of the province in California are, from south to
north, the Trinity, Salmon, and Scott Rivers; these are tributary to
the master stream, the Klamath River; both the Trinity and Salmon
Rivers have tributaries which are themselves major streams. The
Smith River drains parts of the province in northwestern California
and southwestern Oregon. In general, all these streams flow westward,
transverse to the structural grain of the province, although parts of
them and their tributaries have long nearly north-south reaches which
parallel the structural and lithic trends. The southeastern part of the
province is drained by the south-flowing Sacramento River and its
tributaries. In Oregon the principal stream, cutting across the province from east to west, is the Rogue River, with its principal tributaries, the Applegate and Illinois Rivers. The rugged terrain in southwestern Josephine County is drained by the shorter, Chetco River
and its tributaries. Streams in the northernmost part of the province
feed into the South Fork of the Umpqua River and its major
tributary, Cow Creek.

The region is sparsely settled; the principal towns from south to north are Redding, Weaverville, Dunsmuir, and Yreka, in California and Ashland, Medford, Grants Pass, and Canyonville in Oregon. Other smaller settlements are scattered throughout the province. The main road is Interstate 5, which follows the eastern part of the province in California and continues as far as Medford, Oreg., where it turns westward and cuts diagonally across the province to Grants Pass and thence northward through the northern part of the area. U.S. 101 runs a few miles west of the province. Two main east-west roads cross the province in California and connect U.S. 101 to Interstate 5. In the south, California Route 299 connects Redding with Eureka. In the north, California Route 96 joins Interstate 5 near Yreka and continues along the Klamath River to connect with California Route 299 at Willow Creek. U.S. 199 crosses the west-central part of the province between Crescent City, Calif., on U.S. 101, and Grants Pass, Oreg., on Interstate 5. Several good subsidiary routes connect with these main thoroughfares and in turn are joined by private, county, Forest Service, and Bureau of Land Management roads, many of which are used as logging roads; thus most of the region is reasonably accessible. Many of the old gold mines, however, can be reached only by poor or unmaintained trails.

GEOLOGY

The rocks of the Klamath Mountains geologic province belong in two major categories: (1) subjacent rocks older than the Nevadan (Late Jurassic) orogeny; and (2) post-Nevadan superjacent rocks. The subjacent rocks are divided further into pre-Nevadan basement rocks composed of eugeosynclinal sedimentary and volcanic rocks and their metamorphic derivatives, and pre-Nevadan plutonic rocks. Isolated patches of superjacent rocks unconformably overlie the subjacent units and completely surround the province itself.

SUBJACENT ROCKS

A major contribution to the understanding of Klamath Mountains geology was made by Irwin (1960b, p. 16–30; 1966, p. 21–25) who divided the subjacent pre-Nevadan basement rocks into four major units: (1) the eastern Klamath belt, (2) the central metamorphic belt, (3) the western Paleozoic and Triassic belt, and (4) the western Jurassic belt. Davis (1966, p. 39–46) termed these belts subprovinces and called the eastern Klamath belt the eastern Paleozoic subprovince; however, the eastern Klamath belt also includes rocks of Triassic and Jurassic age. The fourfold subdivision of the basement rocks is shown on the accompanying geologic map (pl. 1).

EASTERN KLAMATH BELT

The eastern Klamath belt includes rocks that range in age from Ordovician(?) to Jurassic. Rocks of Ordovician(?) and Silurian age form an elongate belt on the east side of the province south of Yreka. A large area occupied by strata ranging in age from Devonian to Jurassic lies in the southeastern part of the province north of Redding. Both areas include lithologies typical of a eugeosynclinal environment of deposition—that is, graywacke, sandstone, shale and mudstone, chert and chert pebble conglomerate, impure limestone, and a wide variety of volcanic rocks including greenstone, pillow lavas, volcanic breccias and pyroclastics of basaltic composition, spilite and keratophyre flows and pyroclastics, and andesitic flows and tuffs. Strata of the eastern Klamath belt are estimated to have an aggregate thickness of 40,000–50,000 feet (Irwin, 1966, p. 21).

No attempt is made here to describe all the formations of the eastern Klamath belt. Only those units in which gold deposits are known will be discussed. For a more comprehensive review the reader is referred to articles by Kinkel, Hall and Albers (1956), Albers and Robertson (1961), Albers (1964), Churkin and Langenheim (1960), Churkin (1965), Diller (1906), Sanborn (1960), and Wells, Walker, and Merriam (1959).

DUZEL AND GAZELLE FORMATIONS

Rocks of Ordovician(?) and Silurian age in the eastern Klamath belt occupy an elongate area south of Yreka. These rocks are separated from the main area of the eastern Klamath belt by a large ultramafic body which is intruded by gabbroic and granitic plutons.

Diller (1886) and Hershey (1901) first described these rocks and assigned them to the Devonian, assuming a correlation with the Kennett Formation. Later, Wells, Walker and Merriam (1959), on the basis of reconnaissance studies, divided the rocks into two units, the Duzel and Gazelle Formations, and assigned them to the Upper Ordovician(?) and Upper Silurian, respectively. Subsequent more detailed work by Churkin and Langenheim (1960), and Churkin (1965) elucidated the stratigraphy and structure of the Gazelle Formation in a limited area in the eastern part of the region and confirmed its Silurian (Early?, Middle, and Late? Silurian) age. Romey (1962) described the lithology of the Duzel and Gazelle Formations in the approximate southern fourth of the area.

Rocks assigned by Wells, Walker, and Merriam (1959) to the Duzel Formation include phyllitic graywacke with interbedded lenticular limestone and associated chert beds. According to Romey (1962), the most abundant rock types in the Duzel of the southern part of the area are wackes, mudstones, and semischists. He also recognized a massive chert unit. A Late Ordovician(?) age has been assigned (Wells and others, 1959, p. 646) to a large and distinctive coral and brachiopod fauna collected from limestones in the Duzel Formation.

The Gazelle Formation is composed of shale, volcanic graywacke, quartz arenite, chert conglomerate, bedded chert, and limestone. Its main difference from the Duzel Formation seems to be its lack of phyllitic or schistose graywackes. The limestones of the Gazelle have yielded an abundant fossil fauna which Merriam (1961) regarded as of Silurian and Early Devonian age. Graptolites from shale and siltstone indicate that the formation is at least as old as Middle Silurian (Churkin, 1965).

The nature of the contact between the Duzel and Gazelle Formations is essentially unknown; similarly, with local exceptions, the internal stratigraphy of the formations is also unknown. In the southeastern part of the area, phyllitic rocks assigned to the Duzel Formation are thrust over the Gazelle Formation (Wells and others, 1959; Churkin and Langenheim, 1960). Much more mapping is needed to resolve the complex structure of this area of Paleozoic rocks.

COPLEY GREENSTONE AND BRAGDON FORMATION

The Copley Greenstone and Bragdon Formation, which occur together in the southeastern segment of the eastern Klamath belt, are hosts of major gold lode deposits, particularly at or near their mutual contact.

The Copley Greenstone (previously called the Clear Creek Greenstone by Hershey, 1901, p. 233–236; the Copley Meta-andesite by Diller, 1906; and renamed the Copley Greenstone by Kinkel and Albers, 1951, p. 4) is regarded as Early Devonian (Poole and others, 1967, p. 899). It is unfossiliferous but conformably underlies the Balaklala Rhyolite, the upper part of which has been dated as Middle Devonian. It is composed of interlayered volcanic flows, tuffs, agglomerate, and a few thin layers of tuffaceous shale and black shale of small areal extent (Kinkel and others, 1956, p. 10). The igneous rocks include keratophyre, spilite, and albite diabase, some andesite, and some metagabbro (Kinkel and others, 1956, p. 11). Low-grade regional metamorphism of the greenschist facies has affected the mafic volcanic rocks of the Copley Greenstone.

Northwest of Redding the Copley Greenstone is separated from the Bragdon Formation by as much as 3,500 feet of the Balaklala Rhyolite (Kinkel and others, 1956). The Balaklala Rhyolite interfingers with greenstone of the Copley and also occurs as intrusive dikes and sheets within the Copley. Elsewhere in the southeastern part of the province the Balaklala is absent, and the Copley is directly overlain by the Bragdon. The Balaklala is composed of felsic siliceous flows, flow breccias, and pyroclastics which have the composition of keratophyre. The Balaklala is economically important because it is the host for replacement bodies of massive copper-zinc sulfidè ore which have yielded subordinate amounts of gold and silver.

At a few places in the southeastern part of the province, a thin discontinuous cherty shale unit containing some limestone, the Kennett Formation, lies above the Balaklala Rhyolite; in some places it rests directly on the Copley Greenstone. The age of this formation has been established as Middle Devonian on the basis of several fossil collections (Kinkel and others, 1956, p. 37–38).

The Bragdon Formation is one of the most extensive lithologic units of the eastern Klamath belt. It is a rather uniform sequence of dark shale and siltstone with relatively minor amounts of interbedded sandstone and conglomerate. Its thickness is uncertain. Diller (1906) estimated 6,000 feet, but its widespread distribution suggests that the Bragdon either is much thicker or is repeated by folding and faulting. The age of the Bragdon Formation is uncertain, though it is generally regarded as Mississippian on the basis of a few fossils (Diller, 1906). Albers and Robertson (1961, p. 18) suggested that it may occupy a stratigraphic position between the Middle Devonian Kennett Formation and the Mississippian and Lower Pennsylvanian Baird Formation. At many places in the southeastern Klamath Mountains the Kennett Formation is absent, and the Bragdon rests on the Balaklala Rhyolite or the Copley Greenstone.

On the basis of detailed mapping in the French Gulch quadrangle, Albers (1964, 1965) and Albers, Kinkel, Drake, and Irwin (1964) concluded that the contact between the Bragdon and underlying formations is a thrust fault, which was named the Spring Creek thrust. Reconnaissance studies suggest that other small areas of Copley Greenstone surrounded by Bragdon Formation northwest of the French Gulch–Deadwood district are windows in the thrust sheet. Many of the principal gold deposits in the French Gulch–Deadwood area are (1) in the Bragdon Formation above the contact, (2) in veins along the contact between the Bragdon and Copley, or (3) in veins in tabular bodies of quartz porphyry that were intruded along the contact between the Copley and Bragdon Formations.

CENTRAL METAMORPHIC BELT

The central metamorphic belt lies west of the eastern Klamath belt in a great arc extending from southwest of Redding, Calif., to approximately Etna, Calif. Serpentinized ultramafic rocks lie between rocks of the central metamorphic belt and the eastern Klamath belt along its entire east boundary. Its west boundary, with rocks of the western Paleozoic and Triassic belt, is an eastward-dipping thrust fault. Two main units—the Salmon Hornblende Schist and the Abrams Mica Schist—make up the central metamorphic belt.

SALMON AND ABRAMS FORMATIONS

The Salmon Hornblende Schist and Abrams Mica Schist are co-extensive formations in the southern part of the central metamorphic belt. They were originally defined by Hershey (1901) and subsequently studied and redefined by Irwin (1960a, b; 1966), Davis and Lipman (1962), and Davis, Holdaway, Lipman, and Romey (1965).

The Salmon Hornblende Schist is a rather uniform fine-grained, well-foliated hornblende-epidote-albite schist, probably formed by metamorphism of basic volcanic rocks. The Abrams Mica Schist, composed predominantly of metasedimentary rocks, includes quartz-mica schists, calcareous schists and impure marble, and intercalated amphibolite. The Abrams Mica Schist, as used here, differs from Hershey's (1901) original definition in that it excludes rocks that were separated from it and named the Stuart Fork Formation by Davis and Lipman (1962).

The Salmon Hornblende Schist is overlain by the Abrams Mica Schist, and both have a similar metamorphic tectonic history. They were regionally metamorphosed under conditions of the upper greenschist and almandine-amphibolite facies and subsequently underwent retrogressive metamorphism under conditions of the lower greenschist facies (Davis and others, 1965, p. 950–951; Davis, 1966, p. 41). Together, the Salmon and Abrams Formations constitute a folded thrust plate that overrides rocks of the western Palezoic and Triastic belt and that in turn is rooted to the east beneath an ultramafic sheet and the plate of eastern Klamath belt rocks (Irwin and Lipman, 1962; Davis, 1965; Irwin, 1966).

The Salmon and Abrams Formations were considered to be pre-Devonian by Hershey (1901); Diller (1922) and Hinds (1932) regarded them as pre-Silurian, probably Precambrian. Irwin (1960b, p. 20), however, suggested that they might be metamorphosed equivalents of Palezoic rocks in the adjacent eastern Klamath belt. Valid stratigraphic evidence of their age is lacking; however, potassium-argon ages ranging from 270 to 329 m.y (million years) for hornblende and muscovite from the schists in the southern part of the belt suggest that the age of metamorphism was Carboniferous (Lanphere and Irwin, 1965). More recently, rubidium-strontium techniques indicated an isotopic age of 380 m.y. (Devonian) for the initial metamorphism (Lanphere and others, 1968).

WESTERN PALEOZOIC AND TRIASSIC BELT

The western Paleozoic and Triassic belt is the most extensive subunit in the Klamath Mountains province. It constitutes a broad, continuous, north-south-trending belt throughout the entire length of the province. It includes abundant fine-grained clastic sediments, rhythmically bedded chert, mafic volcanic rocks, and scattered lenticular bodies of crystalline limestone or marble. Throughout much of the belt the rocks have been regionally metamorphosed under conditions of the lower greenschist facies, expressed by alteration of the original mafic volcanic rocks to greenstones, development of cleavage in argillaceous rocks, and recrystallization of limestone to marble.

Structural and stratigraphic relationships in these rocks are poorly known, and meaningful division into formations has not been accomplished, although formal names have been given to rocks of this belt at several places. In California these divisions are the blue chert and lower slate series of Hershey (1901), the Chanchelulla Formation of Hinds (1932), and the Grayback Formation of Maxson (1933). Similar rocks in southwestern Oregon are called the Applegate Group (Wells and others, 1949, p. 3–4). Fossils are very scarce, so the age of the rocks is little known. Fossils from a few scattered localities have been identified, ranging in age from Early Permian to Late Triassic (Irwin, 1966, p. 21–22), although collections made during some of the early reconnaissance work were initially considered to range from Devonian to Carboniferous (Diller, 1903, p. 344–346; 1914, p. 15–16). Correlation of these rocks with strata of the eastern Klamath belt has not been suggested, although if age designations are correct, some of them are cœval. Presumably the western Paleozoic and Triassic belt represents a different, perhaps seaward, facies of part of the eastern Klamath belt.

STUART FORK FORMATION OF DAVIS AND LIPMAN (1962)

Davis and Lipman (1962) applied the name Stuart Fork Formation to metamorphic rocks that underlie the Salmon Hornblende Schist and that are exposed in the cores of isoclinal antiforms in the south-central and central parts of the central metamorphic belt. Previously these rocks were included in the Abrams Mica Schist. Similar rocks are essentially continuous northward beyond the area of Salmon and Abrams Formations, constituting a distinctive unit along the east border of the western Paleozoic and Triassic belt.

The Stuart Fork Formation is composed chiefly of complexly folded phyllitic to schistose micaceous quartzites and some interbedded graphitic quartz-mica phyllites, both of metasedimentary origin, and minor amounts of fine-grained greenstones and foliated actinolitic schists that interfinger with the silaceous rocks. Scattered lenses of marble occur throughout the formation. Metamorphism of the Stuart Fork Formation was under conditions of the lower greenschist metamorphic facies. Locally, however, lawsonite-glaucophane-bearing schists derived from greenstones and associated metavolcanic rocks are indicative of the higher pressure blueschist metamorphic facies.

Isotopic ages of the Stuart Fork Formation (Lanphere and others, 1968) range from 133 to 158 m.y. and suggest that metamorphism occurred during the Jurassic and therefore later than metamorphism of the Salmon and Abrams Formations. The distribution of the Stuart Fork along the boundary between the central metamorphic belt and the western Paleozoic and Triassic belt suggests that metamorphism was localized along the sole of the fault during thrusting of the Salmon and Abrams Formations over rocks of the western Paleozoic and Triassic belt during the Nevadan orogeny (Lanphere and others, 1968). The age of the parental rocks is somewhat doubtful, but they are tentatively correlated with rocks of the western Paleozoic and Triassic terrane (Davis and others, 1965, p. 942; Davis, 1966, p. 41; Davis, 1968, p. 915).

AREAS OF HIGH-GRADE METAMORPHISM WITHIN THE BELT

In the central part of the province and at its northeast end, large areas of metamorphic rocks are tentatively regarded as correlative with rocks of the western Paleozoic and Triassic belt but are of a higher metamorphic grade. In northern Jackson and southeastern Douglas Counties, Oreg., metasedimentary rocks of this kind were called the May Creek Formation by Diller and Kay (1924, p. 2); elsewhere they have not been given formal names.

Rocks within these areas are thoroughly recrystallized, well-foliated metasedimentary and metavolcanic rocks of the almandine-amphibolite metamorphic facies. The metasedimentary rocks include biotite-quartz schists (some of them garnetiferous), quartzites, and lenticular bodies of moderately to coarsely crystalline marble. The metavolcanic rocks are represented by fine- to medium-grained amphibolites. In northern California and southern Oregon, rocks of the almandine-amphibolite facies appear to grade into lower grade rocks of the western Paleozoic and Triassic belt (Wells, 1956; Davis,

1966, p. 42; Hotz, 1967). Isotopic analyses of amphibolites from the area northwest of Yreka, Calif., yielded potassium-argon ages of 148 and 146 m.y., suggesting a Jurassic age of metamorphism (Lanphere and others, 1968).

SCHISTS OF CONDREY MOUNTAIN

A subcircular area of schists of the greenschist facies on the California-Oregon State line northwest of Yreka, Calif., is of un-certain affiliation, and the rocks are referred to herein as the schists of Condrey Mountain. They consist of two main types: a commonly graphitic quartz-muscovite schist and an actinolite-chlorite schist. Minor quantities of glaucophane- and stilpnomelane-bearing schists are also present. Mineral assemblages indicate metamorphism under conditions of the lower greenschist facies. The schists are surrounded and structurally overlain along a moderately to steeply dipping reverse fault by rocks of higher metamorphic grade—amphibolite and siliceous schists.

The schists of Condrey Mountain were tentatively correlated with the Abrams Mica Schist by Diller (1914, p. 14). Wells and others (1940) called them older schists of Paleozoic age, and Wells (1956) called them pre-Upper Triassic schists. Rynearson and Smith (1940) referred to them as pre-Mesozoic older metamorphic rocks. Potassium-argon analysis of muscovite from quartz-muscovite schist gives an age of 141 m.y., suggesting a Late Jurassic age of metamorphism. The schists have no clear lithologic correlatives among units of known stratigraphic position and age. Their lithology suggests that they were derived from a sequence of pelitic sedimentary rocks with some interbedded mafic volcanic rocks. Suitable parental rocks might be some of the Mississippian, Triassic, and Jurassic sedimentary rocks of the eastern Klamath belt. However, the Galice Formation of the western Jurassic belt is also of suitable lithology and, were it not so young (Oxfordian and Kimmeridgian), would seem to be a better choice because of its structural position below rocks of the western Paleozoic and Triassic belt elsewhere. The style of deformation and isotopic age are not incompatible with those of the Stuart Fork Formation of Davis and Lipman (1962).

WESTERN JURASSIC BELT

The western Jurassic belt occupies the west side of the Klamath Mountains geologic province for nearly its entire length. It is bound-ed on the east by rocks of the western Paleozoic and Triassic belt and plutons of serpentinite, gabbro, and granitic rocks. This east contact is a fault along which older rocks were thrust westward over the Jurassic rocks. To the west, rocks of the western Jurassic belt are overthrust on the Franciscan Formation in California and Dothan Formation in Oregon.

Rocks of the western Jurassic belt are the clastic and volcanic rocks of the Galice Formation. In Oregon and the extreme northern part of the belt in California, volcanic rocks in the western part of the belt have been named the Rogue Formation (Wells and Walker; 1953; Cater and Wells, 1953). Clastic rocks of the Galice Formation include slaty dark-gray to black fine-grained pelites, fine- to medium-grained graywacke, and subordinate conglomerate. The volcanic rocks are flows, flow breccias, and tuffs of basaltic, andesitic, and dacitic to rhyolitic composition. Commonly, the rocks are weakly metamorphosed; a slaty cleavage has been developed in the fine-grained sedimentary rocks at many places, and the volcanic rocks are composed of sodic plagioclase, actinolitic amphibole, epidote, and chlorite, although original textures and structures generally are preserved.

A long, narrow zone of metamorphic rocks in the western part of the western Jurassic belt in Oregon borders the east side of a large gabbroic and ultramafic body on the west border of the province. Amphibolite is the dominant rock type in this zone, but some interlayered biotite(-garnet)-quartz schist and micaceous quartzite are also present. Wells and Walker (1953) concluded that these rocks are a more highly metamorphosed facies of the Rogue Formation. They resemble amphibolite-grade rocks previously described in some areas of the western Paleozoic and Triassic belt.

The age of the Galice Formation is well established, on the basis of fossils, as Late Jurassic (Oxfordian and Kimmeridgian). The formation is generally considered to be stratigraphically equivalent to the lithologically similar Mariposa Formation in the western Sierra Nevada.

INTRUSIVE ROCKS

The pre-Nevadan basement rocks are intruded by many ultramafic and granitic plutonic bodies. This plutonism in part preceded and in in part concurred with the Nevadan orogeny (Lanphere and others, 1968).

ULTRAMAFIC AND GABBROIC ROCKS

Ultramafic rocks are abundant in the Klamath Mountains province. They range in composition from pyroxenite to dunite, but peridotite (variety harzburgite) containing smaller masses of dunite is most common. Locally the ultramafic bodies are relatively unserpentinized, but for the most part serpentinization is complete.

In general the outcrop pattern of the ultramafic plutons is elongate parallel to the structural grain of the basement rock; this pattern accentuates the arcuate structure of the Klamath Mountains. Outcrops generally range from about 1 acre to several hundred square miles. The major bodies are continuous for scores of miles and commonly occur along the boundaries between the major lithic belts. This feature is best shown by the Trinity ultramafic pluton (Hinds, 1935), which occurs between rocks of the eastern Klamath belt and the central metamorphic belt in the southeastern part of the province. Another large ultramafic body, the Josephine pluton (Wells and others, 1949), on the west border of the Klamath Mountains province, is continuous for approximately 65 miles and separates rocks of the western Jurassic belt from rocks of the Franciscan and Dothan Formations. Similarly, the contact between the western Paleozoic and Triassic belt and the western Jurassic belt is marked at many places by long, narrow bodies of ultramafic rock. Many linear bodies throughout the province occur along known fault zones, but some are not clearly related to known structural features. Groups of small irregularly shaped bodies possibly are remnants of continuous tabular bodies; some of these remnants are infolded with the rocks they intrude.

Mafic rocks that range from diabase to gabbro commonly accompany the ultramafic rocks. Some intrude the same country rocks that are intruded by the ultramafic rocks. Others are enclosed by, and seem to intrude, the ultramafic rocks; others which are surrounded by peridotite have yielded radiometric ages (Lanphere and others, 1968) which are older than those of the rocks intruded by the ultramafic bodies.

GRANITIC ROCKS

Granitic rocks are widespread and intrude rocks of all four lithic belts. They are, however, most plentiful in the north-south axial part of the province, where they intrude rocks of the central metamorphic and western Paleozoic and Triassic belts. The granitic plutons generally range in size from about 1 mile in diameter to about 100 square miles. The plutons tend to be elongate parallel to the north-south arcuate trend of the province. Most have been examined only cursorily, but some in the central metamorphic belt have been studied in detail by Davis, Lipman, Holdaway, and Romey (Davis, 1963; Davis and others, 1965; Lipman, 1963). These investigators found that some of the plutons have domical internal structures and that one is cylindrical. In general, too, the intrusions are concordant with the structure of the enclosing rocks. The small pluton northwest of Yreka (Hotz, 1967) and the large body on the northwest border of the province in

Oregon are tabular, sill-like bodies. An elongate body known as the Ironside Mountain pluton in western Trinity and eastern Humboldt Counties, Calif., may have a similar geometry.

Quartz diorite is the predominant rock type of the granitic plutons, although diorite and granodiorite are also very common; the total range in composition extends from gabbro to quartz monzonite. Plutons in the south-central part of the province, mainly in the central metamorphic belts, are zoned and commonly have trondhjemitic cores (Davis, 1963; Davis and others, 1965; Lipman, 1963). Other plutons have not been studied in enough detail to detect the presence or absence of similar compositional zoning; however, trondhjemite is known from at least one other intrusion located northeast of Grants Pass, Oreg.

Stratigraphic evidence suggests that plutonism occurred after Late Jurassic time but before Early Cretaceous time. The youngest rocks that are intruded by the granites are Late Jurassic, the Galice Formation. The oldest superjacent rocks are the Early Cretaceous (Valanginian) strata that lie unconformably on the eroded Shasta Bally pluton southwest of Redding. These Lower Cretaceous rocks are part of a conformable sequence underlain by Upper Jurassic (Tithonian) strata a few miles to the south (Irwin, 1966, p. 28). Upper Cretaceous strata rest unconformably upon granitic rocks north of Yreka, Calif., and southwest of Medford, Oreg.

Isotopic data, using potassium-argon methods, have been obtained for several of the plutons in California (Curtis and others, 1958; Davis and others, 1965, p. 963; Lanphere and others, 1968) and Oregon (Lanphere, 1969). These data indicate that the granitic plutons fall into four different age groups (Lanphere and others, 1968). The oldest pluton, the Pit River stock north of Redding, Calif., has a minimum age of 246 m.y. (Permian). The other episodes of plutonism fall within the following ranges (in million years): 165–167, 145–155, and 127–140. Distribution of the four groups is illustrated in figure 2.

MINOR INTRUSIVE BODIES

The subjacent rocks are intruded by dikes and sills that range from dark and quartz-free rocks to light-colored quartz-bearing rocks. The dark rocks include diabase, metagabbro, and mafic rocks called lamprophyre in the old reports. Rocks of intermediate composition include andesite, andesite porphyry, and diorite porphyry. Light-colored siliceous rocks include quartz porphyry, dacite porphyry, and aplite.

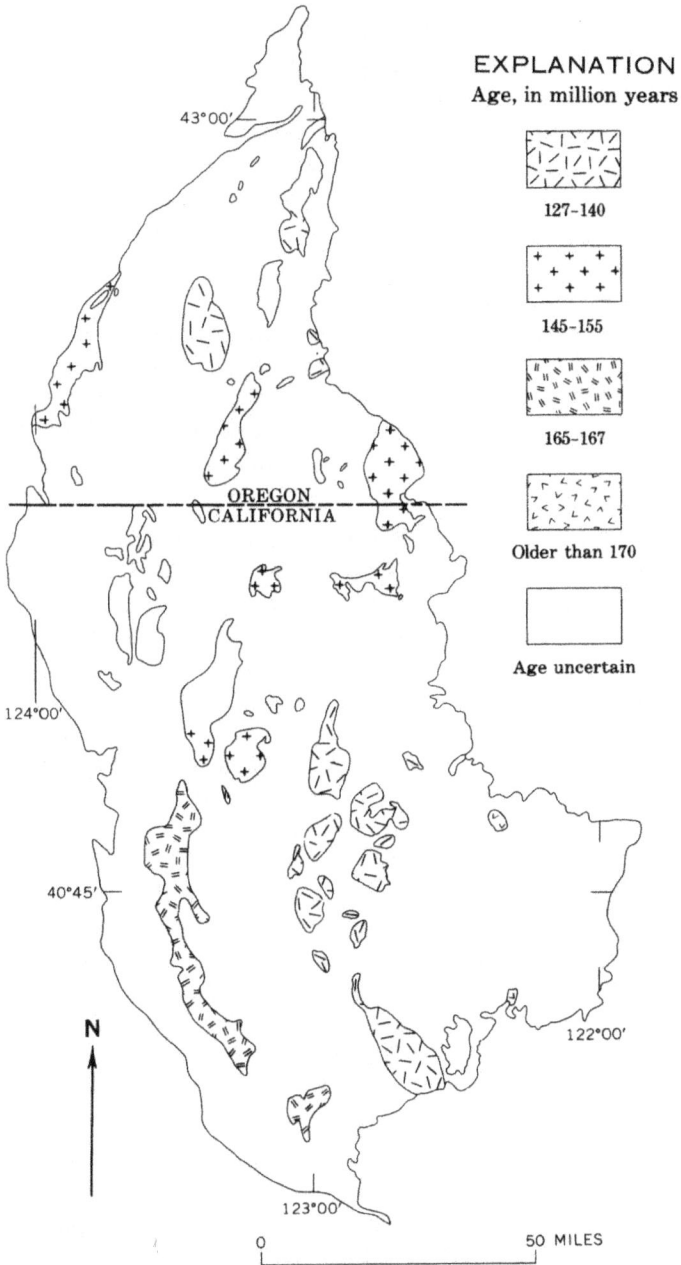

FIGURE 2.—Isotopic ages of granitic plutons in the Klamath Mountains.

Certain of the minor intrusive bodies commonly crop out near mineralized areas or are exposed underground in many of the mines; these were described in many published reports on the old mines and mining districts. One such intrusive rock was called "birdseye porphyry"; this term was used by prospectors and miners to describe light-gray porphyritic rocks with conspicuous phenocrysts of plagioclase feldspar in a fine-grained equigranular to aphanitic groundmass. The name is derived from the appearance of weathered zoned plagioclase phenocrysts which resemble the eye of a bird. In addition to plagioclase, phenocrysts of hornblende, biotite, and rounded lobate corroded quartz are common. The composition of the dikes ranges from nonquartzose porphyry to varieties in which quartz phenocrysts are abundant. Birdseye porphyry commonly is fresh to only slightly altered. Alteration, probably deuteric, results in partial clouding and sericitization of the feldspar and chloritization of hornblende and biotite.

Some dark mafic dikes are present in most of the mineralized areas, but they are not evenly distributed among the areas; for example, they are common in the Trinity Center district, but are very rare in the Liberty and Yreka–Fort Jones districts. In many of the old reports these dark dikes are called lamphrophyres, a term that seemingly was indiscriminately used for any dark fine-grained rock. Knopf (1936, p. 1748) defined lamprophyre as "mesocratic or melanocratic rocks carrying solely ferromagnesian phenocrysts in an aphanitic or microgranular groundmass, and in which the ferromagnesian minerals in the groundmass show notable idiomorphism." Some of the dark dikes described in the literature on the Klamath region and examined by me conform to this definition. They are fine-grained porphyritic rocks with phenocrysts of hornblende, less commonly of biotite, or both. Feldspar phenocrysts are conspicuously absent, feldspar being restricted to the groundmass in which the dark ferromagnesian minerals are also idiomorphic. Other dark mafic nonporphyritic dikes in the Klamath Mountains province are more properly classed as basalt, diabase, or metagabbro.

The most common dike rocks are medium-gray fine-grained nonporphyritic to slightly porphyritic andesites. The phenocrysts are mainly plagioclase, with minor pale-green to brownish-green amphibole, which commonly is altered. The fine-grained groundmass is composed of euhedral to subhedral calcic oligoclase to sodic andesine and euhedral to subhedral pale-brown to green hornblende. Microscopic examination shows that anhedral quartz is also an essential constituent of several specimens, so many of these apparently andesitic dikes actually should be classified as dacites. These dike rocks commonly are partially altered, probably in part deuterically, and contain abundant interstitial chlorite.

There have been few field observations of crosscutting relations among the minor intrusives, so their relative ages are inadequately known. Most of the bodies probably are younger than the granitic intrusions (Albers, 1964, table 1). Some of the silicic bodies whose compositions are similar to those of the granitic intrusives may be hypabyssal equivalents of the granitic rocks. According to several reports (MacDonald, 1913, p. 9; Kinkel and others, 1956, p. 53; Albers, 1964, p. J51), the lamprophyre dikes are the youngest of the minor intrusive bodies.

SUPERJACENT ROCKS

Superjacent rocks in the Klamath Mountains province rest unconformably on the pre-Nevadan basement rocks and the plutonic intrusions, and are of Cretaceous, Tertiary, and Quaternary age. Excepting those of Quaternary age, superjacent deposits within the province are not plentiful, although scattered remnants suggest that they were once more extensive.

Cretaceous rocks overlap the older, subjacent rocks at the north, northeast, and southeast borders of the province; several isolated outcrops occur within the province. These rocks are marine deposits typically composed of firmly cemented well-bedded conglomerate, sandstone, and mudstone. Locally they are abundantly fossiliferous. At the north end of the province, Upper Jurassic and Lower Cretaceous rocks of the Myrtle Group rest unconformably on the Upper Jurassic Galice and Rogue Formations. An isolated outcrop in Oregon near the State line approximately 30 miles southwest of Grants Pass is correlative with the upper part of the Myrtle Group. A larger area of Cretaceous rocks northeast of Grants Pass has been designated uppermost Lower Cretaceous (Albian) by Jones (1960). The beds that overlap the subjacent terrane and occur as small isolated remnants in the northeastern part of the province near Yreka, Calif., and Medford, Oreg., are of Late Cretaceous age (Peck and others, 1956) and are correlated with the Hornbrook Formation. The small isolated patches of Cretaceous rocks in the southeastern part of the Klamath Mountains province are correlative (Murphy and others, 1964) and undoubtedly were formerly continuous with Lower Cretaceous rocks of the Great Valley sequence that overlap the subjacent terrane along the southeast boundary of the province. These strata, composed of several members, have been named the Budden Canyon Formation by Murphy, Peterson, and Rodda (1964).

Sedimentary rocks of Tertiary age flank the north and northeast borders of the Klamath Mountains but rarely crop out within the province. On the west edge of the Klamath Mountains province east of Crescent City, Calif., there are a few thin remnants of a marine deposit. These strata, named the Wymer Beds by Diller (1902, p. 31–35) and the Wimer Formation by Maxson (1933, p. 134), are composed of friable shale, siltstone, sandstone, and conglomerate. Their fossil content indicates a late Miocene age (Diller, 1902, p. 31–35). Some terrestrial gravel deposits, reportedly gold bearing (Maxson, 1933, p. 140–143; Cater and Wells, 1953, p. 104), which lie east of the Wimer Beds, are of uncertain age but possibly are penecontemporaneous with the Wimer Formation. In the southern part of the province near Weaverville the Oligocene(?) (MacGinitie, 1937) Weaverville Formation (Hinds, 1933, p. 115; Irwin, 1963) is preserved in several downfaulted blocks. The strata, including conglomerate, sandstone, shale, and tuff, were deposited in continental fluviatile and possibly in part lacustrine environments. Carbonized vegetal remains occur throughout the formation, and locally there are beds of lignite several feet thick. The formation is at least 2,000 feet thick and rests unconformably both on the subjacent rocks and superjacent Cretaceous rocks. Small amounts of gold have been recovered from the Weaverville Formation.

Volcanic rocks of Tertiary age belonging to the Cascade province overlap the subjacent rocks along the northeast border of the Klamath Mountains province in Oregon and are probably continuous with overlapping volcanics in California. These volcanic rocks belong to the western Cascade subdivision and are of Eocene and Miocene age. Near Medford, nonmarine sedimentary rocks of Eocene age that also overlap the subjacent rocks underlie the volcanic rocks. Farther south, on the southeast border of the Klamath Mountains, the overlapping volcanic rocks are Pliocene and Pleistocene in age.

The youngest superjacent rocks are unconsolidated to semiconsolidated fluviatile deposits that occur as valley fill, remnants of stream terraces, and modern sands and gravels along the present streams of the Klamath Mountains. Much of the gold produced from the region has come from placer deposits in these sediments.

STRUCTURAL RELATIONS

Many structural details of the Klamath Mountains province are uncertain or unknown, but in recent years, largely owing to the perceptive syntheses of Irwin (1960b, 1966; Irwin and Lipman, 1962) and mapping by Irwin (1963) and Davis, Holdaway, Lipman, and Romey (1965; Davis, 1968), some general features of the structural relationships have become apparent. The best established structural feature is the parallel arrangement of north-south-trending lithic units or belts, which curve in an arcuate fashion convex to the west. The boundaries between the belts either are faulted or are obscured by elongate bodies of ultramafic rock and local granitic plutons which presumably occupy zones of faulting. Discovery of outliers of adjacent lithologic belts in all except the eastern Klamath belt (Irwin, 1960b) and recognition of a probable klippe of eastern Klamath rocks resting on rocks of the central metamorphic belt near Weaverville (Irwin, 1963) led to the hypothesis (Irwin and Lipman, 1962; Irwin, 1964, 1966) of regional westward thrusting of the major lithologic divisions. More detailed studies in the south-central Klamath Mountains (Davis and others, 1965; Davis, 1968) have resulted in additional field data in support of the thrust-fault hypothesis. The lithic belts are believed to be thrust plates, each of which overrode the adjoining westward belt.

In the large plate of eastern Klamath belt rocks in the southeastern part of the province, the stratigraphic succession becomes increasingly younger eastward and is essentially an eastward-dipping sequence, complicated by folding and faulting, much of which is poorly known. A major structural feature in the southern part of the eastern Klamath belt is the Spring Creek thrust (Albers, 1961, 1964), which is the contact between the Bragdon Formation and the underlying Copley, Balaklala, and Kennett Formations. The amount and direction of movement on the Spring Creek thrust are unknown (Albers, 1964, p. J64), but the thrust may underlie a wide area in the western part of the eastern Klamath belt and may be exposed where the underlying rocks can be seen in windows in the Bragdon Formation. This thrust, modified by later faulting and folding, was a prime factor in the localization of some of the major gold deposits in the French Gulch–Deadwood district.

The area south of Yreka, underlain by the Duzel and Gazelle Formations, is considered to be part of the eastern Klamath belt, but the relation of these formations to rocks in the southeastern part of the province is unknown. The internal structure of the block containing the Duzel and Gazelle is poorly known; but the block is complexly folded, and Churkin and Langenheim (1960) have shown that there is at least one thrust fault on which rocks assigned to the Ordovician(?) Duzel Formation have been thrust over the Gazelle Formation of Silurian age.

Klippen of the eastern Klamath belt overlie rocks of the central metamorphic belt near Weaverville and Cecilville. The klippe near Weaverville is composed of slate of the Bragdon formation (Irwin, 1963). Davis (1968, p. 917) concluded that rocks of the klippe near

Cecilville belong to the Duzel Formation. A sheet of serpentinite underlies both klippen and separates them from the underlying metamorphic rocks.

The large ultramafic body in the eastern part of the province and the serpentinite that everywhere separates rocks of the eastern Klamath belt from the central metamorphic belt are believed to be part of a sheetlike ultramafic pluton that occupies the thrust fault between the two belts (Irwin and Lipman, 1962; Irwin, 1966, p. 31; LaFehr, 1966; Davis, 1968, p. 924).

The gross structure of the central metamorphic belt is essentially that of a complex asymmetrical antiform (Davis and others, 1965; Irwin, 1966, p. 33, fig. 6). In the southern part of the belt, rocks of the Stuart Fork Formation of Davis and Lipman (1962), underlying the Salmon Hornblende Schist, are exposed as fensters in the core of the antiform. Because the Stuart Fork is believed to be correlative with rocks of the western Paleozoic and Triassic belt, and therefore younger than the overlying Salmon Hornblende Schist, the contact between the Stuart Fork and the Salmon Hornblende Schist is considered to be a folded thrust fault on which the Salmon and Abrams Formations are thrust over rocks of the western Paleozoic and Triassic belt. Farther north the Salmon and Abrams are progressively cut out by ultramafic rocks and the thrust sheet of the eastern Klamath belt. The west boundary of the central metamorphic belt with rocks of the western Paleozoic and Triassic belt is probably a thrust fault which has been modified locally by high-angle normal faulting (Irwin, 1966, p. 33; Davis, 1966, fig. 1; Davis, 1968).

The structure of the western Paleozoic and Triassic belt is also poorly known. In general, lithologic units strike north in conformity with the regional arcuate trend of the belt. Most commonly bedding is moderately to steeply inclined eastward, although locally it dips in the opposite direction. Details are lacking in most places, but it is known that the rocks are folded, and it appears that many of the folds are tight and complex with eastward-dipping axial planes.

West and northwest of Yreka and in the vicinity of Seiad Valley and adjoining areas in southern Oregon, the siliceous schists and amphibolite, which are thought to be recrystallized equivalents of western Paleozoic and Triassic belt rocks, have undergone complex, probably recumbent, folding and penetrative shearing. Northwest of Yreka, tightly folded, highly sheared schists of uncertain affiliation are exposed in a window through a thrust plate that has been breached by erosion of a broad north-south-oriented antiform in rocks of the western Paleozoic and Triassic belt and their metamorphosed equivalents. The trace of the exposed thrust fault is a prominent feature of the map in this area.

The boundary of the western Paleozoic and Triassic belt with the western Jurassic belt is a well-defined thrust fault south of Grants Pass, Oreg. (Wells and others, 1949, p. 3). Elsewhere, however, the two belts are separated by elongate bodies of ultramafic rock or by granitic plutons. Yet, the entire contact is believed to be a major eastward-dipping thrust fault which afforded a favorable site for the emplacement of the ultramafic bodies and localized the emplacement of granitic plutons. In the southwestern part of the province, rocks of the western Paleozoic and Triassic belt which crop out within the western Jurassic belt are probably klippen of the eroded thrust sheet.

Sedimentary rocks of the western Jurassic belt commonly have a well-developed slaty cleavage that in most places dips moderately to steeply eastward, subparallel to the moderately dipping axial planes of asymmetric to isoclinal folds. The folds trend north to northeast. High-angle faults, many of which undoubtedly have reverse displacements, parallel the strike of bedding. Some of these are major zones of dislocation that contain narrow slivers of sheared serpentinite.

The contact between rocks of the western Jurassic belt and the Franciscan and Dothan Formations of the Coast Range province is believed to be a fault of regional extent. In the southwestern part of the Klamath Mountains province, eastward-dipping rocks of the Galice Formation overlie an eastward-dipping belt of schist that grades downward into rocks of the Franciscan Formation. Formation of the schist has been attributed to cataclasis and recrystallization beneath the sole of an eastward-dipping thrust fault (Blake and others, 1967). A narrow zone of serpentinite is commonly observed at the contact, and farther north a narrow belt of serpentinite between the Galice and Franciscan Formations is continuous in Oregon with the great Josephine ultramafic body. Northwest of Grants Pass, Oreg., the contact between rocks of the western Jurassic belt and the Dothan Formation is clearly a fault which was observed to dip moderately eastward. The present interpretation is that rocks of the western Jurassic belt were thrust westward over rocks of the Coast Range province, probably during the Late Cretaceous (Irwin, 1964; Blake and others, 1967).

GOLD DEPOSITS

The first reported gold discovery (Raymond, 1874, p. 143) in the Klamath Mountains was made on Clear Creek, Shasta County, Calif., by Maj. Pearson B. Reading in the fall of 1848, soon after Marshall's discovery at Sutter's Mill in the Sierra Nevada. The earliest discoveries in southern Oregon were made in 1852 when placer gold was found on Rich Gulch, a tributary of Jackson Creek, and on Josephine Creek (Winchell, 1914, p. 23–24). The first lode mining began in 1852 at the Washington mine in Shasta County, Calif. (Ferguson, 1914, p. 33). In 1966 there were more than a thousand known lode mines and prospects in the Klamath Mountains of Oregon and California, but only one or two were operating.

The Klamath Mountains province never attained the importance of the Sierra Nevada as a gold-producing region; however, it ranks second in the amount of gold mined in California. It is difficult to determine accurately the amount produced, but it is estimated that the Klamath Mountains region has yielded, from placers and lodes, more than 7 million ounces of gold valued at over $165 million. (For California, see Irwin, 1960b, p. 64; U.S. Bur. Mines, 1955–65. For Oregon, see Libbey, 1963, p. 108.) Separate production data for lodes and placers are not available prior to 1903 and after 1953 for Oregon and 1956 for California, but from 1903 to 1960 approximately 1.5 million ounces were mined from lodes (figs. 3, 4) and 2.2 million ounces from placers (U.S. Bur. Mines, 1927–33; 1933; 1934–67; U.S. Geol. Survey, 1904-26). To arrive at these figures, the production data used were for Shasta, Siskiyou, and Trinity Counties, Calif., and Curry, Douglas, Jackson, and Josephine Counties, Oreg.

GENERAL CHARACTER

Many of the following data on the lode gold deposits of the Klamath Mountains have been obtained from the literature, because workings of most of the principal mines are now inaccessible.

GOLD-QUARTZ VEINS

Most of the lode deposits are described as fissure veins, meaning fractures, usually faults, into which vein material has been introduced.

In general the mineralogy of the veins is simple. The major constituent is quartz, which is usually milky white. Commonly it contains drusy cavities, and some quartz veinlets have a median cavity into which small crystals project. Calcite commonly accompanies quartz in minor amounts. Other nonmetallic gangue minerals include sericite and, rarely, albite.

Pyrite is the most widespread metallic gangue mineral. It occurs in quartz, in wallrock inclusions, and in the wallrock next to the veins; it forms individual crystals and veinlets. Other metallic sulfide minerals which may be present in subordinate amounts are arsenopyrite,

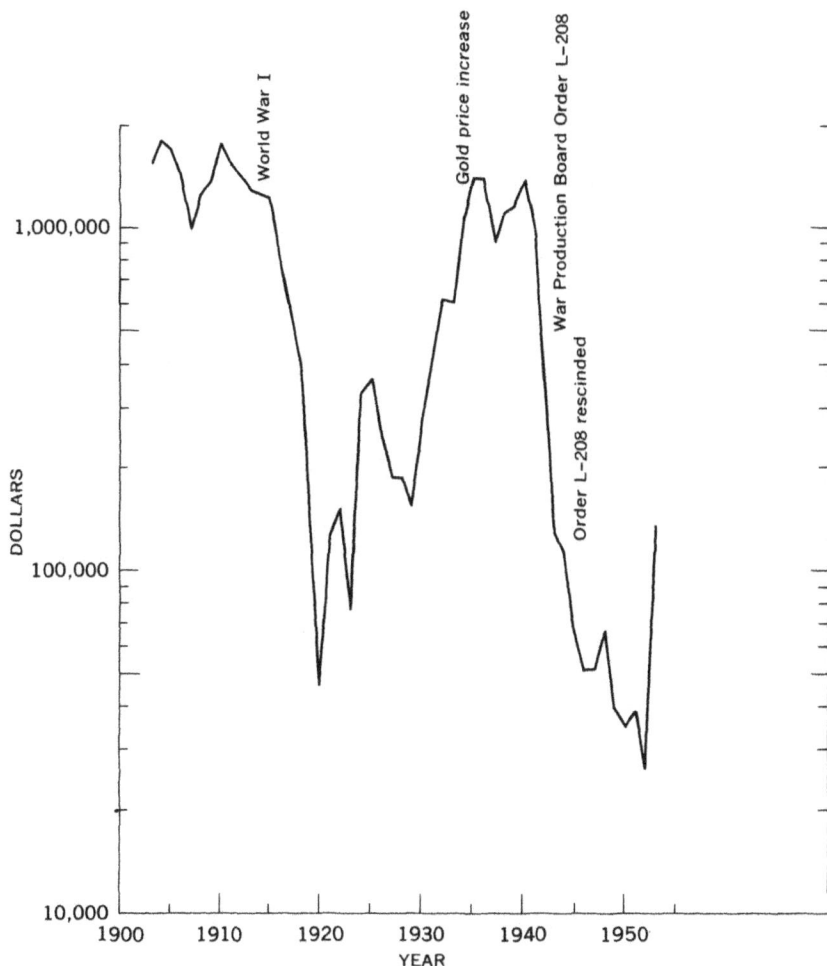

FIGURE 3.—Lode gold production, in dollars, from the Klamath Mountains, 1903 to 1953.

galena, and sphalerite. Still less common are chalcopyrite, pyrrhotite, and molybdenite. The amount of sulfides in most quartz veins is less than 3 percent but may be as much as 5 percent.

Gold occurs almost exclusively in the veins. It occurs as free metal in quartz and in the sulfide minerals, mainly pyrite. Gold tellurides have been reported from some veins.

Many veins contain wallrock fragments; indeed, in some of the most productive veins wallrock fragments are abundant; veins in slate are commonly ribboned with included fragments, and the quartz is bluish, due to comminuted slate (Ferguson, 1914, p. 59–71).

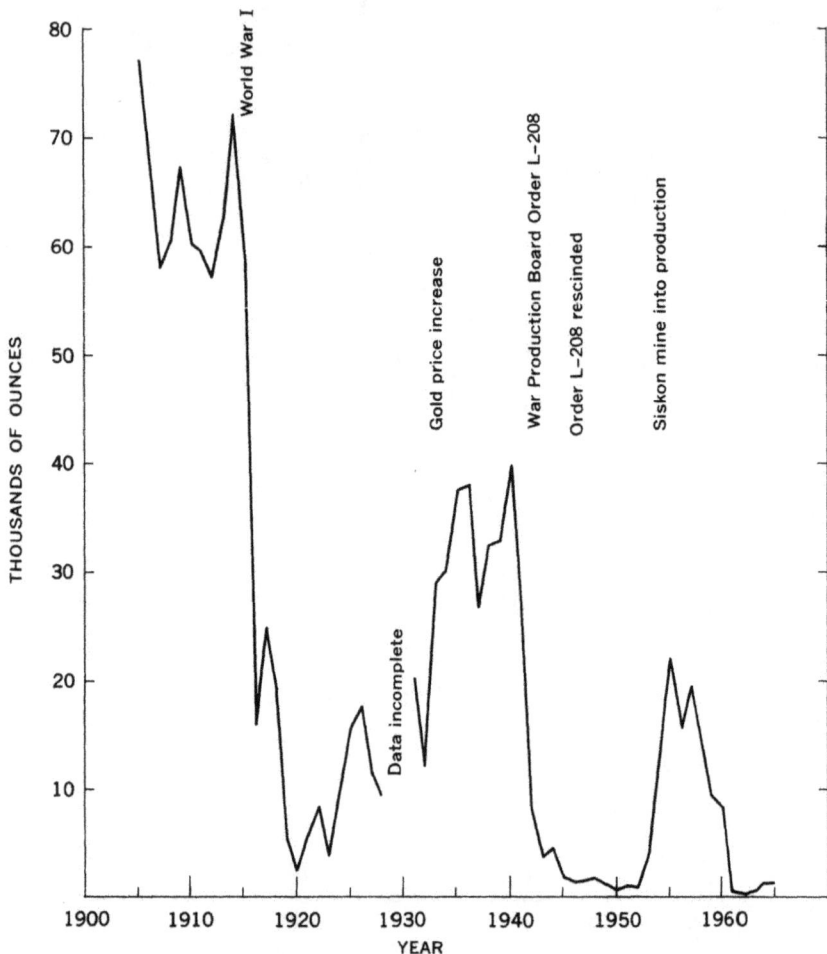

FIGURE 4.—Lode gold production, in ounces, from the Klamath Mountains, 1905 to 1965.

The quartz-filled fractures range from veinlets a fraction of an inch in width to great masses 10 feet or more thick; however, most commonly they are 1 to 5 feet thick. The quartz bodies are more or less lenticular, both horizontally and vertically, so that the veins pinch and swell and may disappear completely. In places the main body of quartz may feather out into several thin nonpersistent stringers. The veins seldom follow a single simple fracture; rather, they occupy complex structures which branch and rejoin, thus enclosing bodies of country rock called horses. Locally veins send off subparallel subsidiary branches that may rapidly terminate or extend long distances.

Commonly a fracture zone contains two or more parallel veins or a series of shorter discontinuous veins arranged en echelon. The course of the veins is generally relatively straight, although in detail the strike is variable, and the dip may vary within several degrees. Most veins are steeply dipping, although some are known to be inclined at relatively low angles. The contacts of many veins with the wallrocks are slickensided owing to postmineralization movement, resulting in clearly defined vein boundaries. In other veins only one contact is faulted while the other is frozen to the wall and sends out veinlets and stringers of quartz into the wallrock. In rare instances neither wall shows postmineralization movement.

The strike length of most of the veins is no more than a few hundred feet. Several of the most productive veins, however, have been followed for 2,000 feet or more. The most persistent of these veins are in the southeastern part of the Klamath Mountains province in the French Gulch–Deadwood district. The vertical depths to which veins extend are much the same as their horizontal dimensions. Most of them have been followed to less than 500 feet, and a few, to over 1,000 feet; the mining depth, of course, is dependent on mining economics rather than on the real persistence of the vein. Ferguson (1914, p. 37) pointed out that the minimum vertical range through which the gold-quartz veins were deposited in the French Gulch district was 3,400 feet, as determined by the difference between the altitude of the highest outcropping veins in the area and the lowest underground point reached by mining. However, no individual vein is known to be this persistent.

The auriferous veins are not uniformly productive throughout their extent. The enriched, economically productive parts of a vein are called ore shoots. Most of the ore shoots have lengths of a few hundred feet and persist in depth for similar distances, although in some of the more productive mines, shoots have been followed in depth for several times their strike length. The pitch (or rake) of the shoots is generally steep but is seldom the same as the dip of the vein and rarely as low as 45°. Commonly the ore shoots are the widest part of the veins.

For many of the mines in the Klamath Mountains, the controls on localization of the ore shoots are poorly known. From observations made by Ferguson (1914), Shenon (1933a), Youngberg (1947), and Albers (1965), the most obvious controls have been structural. Most commonly the shoots occur at the intersection of veins. These vein intersections may be the junction of subparallel veins or at branches of the same vein, the intersection of a main vein with subsidiary veins, or along the junction of veins with opposing dips. Changes in strike and variations in dip of a vein also have been recognized (Youngberg, 1947) as favored sites for ore shoots. In the French Gulch district, ore shoots also occur where veins intersect the contact between shale and porphyry and where they intersect the contact between the Copley Greenstone and slates of the Bragdon Formation (Albers, 1961,1965).

The known gold content of veins throughout the Klamath Mountains has a wide range and, of course, is highly variable within individual veins. The best available data indicate that most mined veins contained 0.1–0.4 ounce per ton. In the most successful mines, except those of the French Gulch–Deadwood district, the grade of ore, not including enriched surficial ore and pockets, was approximately 0.4–0.75 ounce per ton. The ores of the French Gulch–Deadwood district, which were unusually rich, ranged from 0.5 to more than 1 ounce per ton, with as much as 3–6 ounces per ton in some shoots.

Data on the purity of the gold as mined are practically nonexistent; however, Ferguson (1914, p. 71) stated that gold from the Brown Bear mine in the French Gulch–Deadwood district was 840 fine. Some data on the fineness of gold bullion as received at the mint from California mines are available (Leach, 1899), but these figures may reflect changes caused by the treatment of the ore (Knopf, 1929, p. 37). According to the mint report (Leach, 1899, p. 184–187), the fineness ranged from 718 to 950. Gold from more than half the mines ranged from 750 to 850. Nearly one-fourth of the bullion was between 850 and 900 fine.

GOLD-SULFIDE DEPOSITS

A second, less common but nevertheless significant, kind of gold deposit in the Klamath Mountains is designated here as the gold-sulfide type. The Headlight, Siskon, and King Solomon mines are examples of this type. The gold of these veins was recovered not from the sulfides but from their weathered, oxidized outcrop. This type of deposit is perhaps similar to the large sulfide bodies in the southeastern part of the province that contain minor amounts of gold but were mined principally for their copper-zinc content. The latter have been extensively studied and described by Kinkel, Hall, and Albers (1956) and Albers and Robertson (1961).

Unlike the gold-quartz veins which were deposited in open fissures, the gold-sulfide bodies are replacement deposits. Their tabular form, however, suggests that they occupy well-defined fractures or fracture zones. The sulfide minerals are fine-grained pyrite and subordinate amounts of chalcopyrite, sphalerite, and galena. The chief gangue mineral is fine-grained quartz with a microscopic flamboyant or feathery habit (Kinkel and others, 1956, p. 88, fig. 44). At most deposits, such as at the Siskon and Headlight mines, the mineralized zone is a virtually complete replacement of the country rock by massive fine-grained quartz and pyrite. Less commonly, as at the King Solomon and Queen of Sheba mines, the country rock is only partly replaced but heavily charged with sulfides and minor quartz. Gold is very finely divided and is incorporated in the sulfides, mainly pyrite; none is known to occur free in the quartz. Grade of the protore is low, amounting to a few hundredths of an ounce per ton, and the material has not been mined profitably. Weathering, however, has removed sulfides and has concentrated the relatively insoluble gold in ferruginous siliceous gossans. The gold content of such a gossan may be several tenths of an ounce per ton.

POCKET DEPOSITS

An uncertain but probably considerable amount of gold produced from lodes in the Klamath Mountains was taken from small localized concentrations near the surface which are commonly referred to as pocket deposits. Southwestern Oregon, especially, is considered by prospectors and miners to be pocket country. The genesis of these gold pockets is poorly understood.

Gold pockets are small, shallow, relatively high grade deposits of free gold with little gangue. They apparently have little or no lateral or vertical continuity, because little effort was made to explore possible continuations of a pocket once the rich, easily attained gold had been extracted.

In the southeastern Klamath Mountains Hershey (1910) and Ferguson (1914, p. 40–43) described pocket deposits in the Minersville and Whiskeytown districts that occurred at the contact between slate of the Bragdon Formation and the Copley Greenstone, or at an interface between slate and a dike of diorite porphyry. Veinlets of manganiferous quartz and calcite accompany the gold, which is described as occurring in thin sheets in some places. At three localities — the Eldorado, Mad Mule, and Five Pines mines—the pockets are sufficiently large and close together to have warranted rather extensive work along the contact; all the workings are within a few hundred feet of the surface. Most commonly, however, pocket deposits have been mined for only a few feet below the surface and abandoned after the obvious gold had been exhausted; most of the pockets in southwest Oregon were seldom mined to depths of more than 25 feet (Diller, 1914). The yield from individual pockets in the Minersville and Whiskeytown districts was from $2,500 to as much as $45,000 (Ferguson, 1914, p. 56, 74). The most famous pocket mine in southwest Oregon was the Gold Hill pocket, which yielded approximately $700,-

000 worth of gold within a depth of 15 feet (Diller, 1914, p. 45–56). Many other pockets have been mined in the Klamath Mountains from which gold worth a few hundred to several thousands of dollars was extracted before pockets were exhausted and forgotten.

Hershey (1910) concluded that the pocket gold was deposited by gold-bearing meteoric water. He postulated that the gold was dissolved from weathered auriferous pyrite and precipitated by reaction of the solution with carbon in slate of the Bragdon Formation. Ferguson (1914, p. 41–43; 1915) also subscribed to a theory of supergene origin with precipitation being caused by carbon in the black slates. He suggested, however, that acid waters dissolved gold from low-grade manganiferous quartz veins in metavolcanic rocks through the agency of manganese dioxide. After traveling in solution a short distance, the gold was deposited in irregularities along slate–metavolcanic rock contacts where small calcite lenses neutralized the solutions. Studies by Krauskopf (1951) and Cloke and Kelly (1964) support Hershey's and Ferguson's hypotheses. These studies indicate that acid meteoric waters containing Cl^- and MnO_2 can dissolve and transport gold. In the presence of H^+ the MnO_2 oxidizes the gold, which then combines with Cl^- according to the reaction:

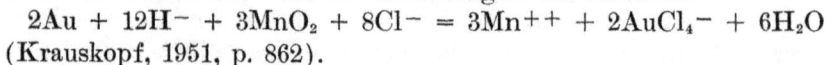

$$2Au + 12H^- + 3MnO_2 + 8Cl^- = 3Mn^{++} + 2AuCl_4^- + 6H_2O$$

(Krauskopf, 1951, p. 862).

Supergene processes may have been responsible for the formation of some of the pocket deposits, but it seems unlikely that the exceedingly rich Gold Hill and Steamboat pockets in Oregon were formed this way. Under certain conditions, where gold precipitation takes place in response to hydrothermal alteration of country rock by vein-forming solutions, it is expectable that rich zones or pockets would occur in conjunction with lean or barren zones (Helgeson and Garrels, 1968). It is also unlikely that gold would be transported a very great distance before neutralization of the solutions would cause precipitation; hence, supergene deposits probably are not far from the original hypogene deposits. The presence of carbonaceous sedimentary rocks may not be necessary to the formation of pocket deposits, because in southwest Oregon gold pockets also occur in areas where there are no carbonaceous rocks (Wells and others, 1949, p. 21). Probably many pocket deposits are irregular and discontinuous shoots along hypogene veins and veinlets which have been enriched by supergene processes.

COUNTRY ROCKS

Although nearly all the rock types that occur in the Klamath Mountains province have been reported as hosts for gold-bearing veins, most of the veins are found in a relatively few kinds of rock. There is no compelling evidence, however, that certain rock types were more favorable than others for the localization of gold-bearing veins. Most commonly the veins occur in metavolcanic rocks, a fact which may be genetically important but, as seems more likely, may reflect the relative preponderance of metavolcanic rocks over other kinds in the Klamath Mountains province. In some areas, argillaceous sedimentary rocks are the favored host. Gold-bearing veins do not usually occur in granitic rocks; however, there are many exceptions to this generalization. Diller (1914, p. 24) commented that many of the gold-bearing veins in southwestern Oregon are near serpentine, but this observation does not apply to the entire province. It seems more likely that both the veins and the ultramafic rocks were localized by the same independent feature—fracturing.

The country rocks of most of the sulfide replacement deposits seem to have been mafic in composition. Some are greenstones and greenschists (Siskon and other mines of the Happy Camp district). For the Headlight property and others in the Trinity Center (old Carrville) district, MacDonald (1913, p. 10) reported that the deposits are replacements of mafic dikes. The King Solomon and Queen of Sheba deposits are unusual in that they are in tactites which were formed by metamorphism of carbonate rocks, probably impure limestones.

The importance of dike rocks to the localization of gold-bearing veins is difficult to assess, although Albers (1961) stated that the great majority of the principal lode deposits in the Klamath Mountains are closely associated with dikes and sills of birdseye porphyry. Certainly in the most productive district, the French Gulch–Deadwood district, there are numerous silicic dikes and sills of birdseye porphyry and quartz porphyry (Albers, 1961, 1965, 1966). In the Trinity Center district mafic basaltic and lamprophyric dikes, which have generally been considered to be ore makers (MacDonald, 1913), are plentiful in addition to birdseye porphyry, and Hershey (1900, p. 88) found an apparent correlation between gold lodes and a diorite porphyrite which he regarded as later than the birdseye porphyry. Similar dikes are found in the other districts in northern California, and they are most plentiful in areas in which mineralization, as indicated by the number of mines and prospects, has been concentrated in relatively restricted districts (for example, the Liberty district south of Sawyers Bar in the Salmon River area). In southern Oregon, however, dike rocks are less commonly found in conjunction with the gold veins, although mafic dikes have been observed at a few places, and Diller (1914, p. 24) noted that "it is not unusual to find rich ores near the contact ... [between greenstones and sedimentary rocks] ... and closely related to dikes which cut them." There seems to be some correlation between the dikes, both mafic and silicic, and auriferous veins, but the relative abundance of similar dikes outside the mineralized areas is unknown because of the lack of detailed geology.

At many places gold-bearing veins are along the contact of dikes with the country rock; less commonly a vein is within a dike. There is no evidence that the dikes were directly responsible for the introduction of metallizing solutions. The dikes are, however, indicative of igneous activity which may have been the source of gold-bearing solutions. Undoubtedly in many places dikes have controlled the location of veins, because their contacts with the country rock, being real or potential surfaces of discontinuity, became channelways for the mineralizing solutions. Hershey (1900, p. 94) arrived at a similar conclusion, suggesting that the observed association of dikes with many of the veins is no proof that the gold was derived from the dikes, but that there is probably an accidental structural connection.

WALLROCK ALTERATION

It is difficult to determine the extent to which the country rock adjacent to the veins has been transformed by the passage of the mineralizing solutions. Ferguson (1914, p. 45) and MacDonald (1913, p. 12) noted that the rocks adjoining the veins in the southern Klamath Mountains show comparatively little alteration directly attributable to the ore-bearing solutions. Diller (1914, p. 25) suggested that wallrock alteration in the southwestern Oregon districts was relatively mild. Observations during the recent investigation tend to substantiate the conclusions of the earlier workers, although most of this new information comes from thin-section studies of specimens from mine dumps and surficial exposures of rocks adjacent to veins at old mine openings.

Where the veins are in slate or argillite, the wallrock is comparatively little altered, although there has been some sericitization and introduction of small amounts of sulfide minerals, chiefly pyrite. Metavolcanic and dike rocks adjacent to veins bear more evidence of hydrothermal alteration. Many specimens contain abundant carbonate, which is only present in minor amounts in unaltered specimens. Most of the carbonate-bearing phase is calcite, but some ankerite was observed. Sericite, with or without calcite, is also common. Many wallrocks also contain chlorite, which probably is in part due to hydrothermal alteration of ferromagnesian minerals, although some is surely of an earlier generation, formed by low-grade regional metamorphism. Dusty and skeletal pseudomorphic replacements of original ferromagnesian silicates by an opaque mineral, probably magne-

tite, are unusual indicators of alteration, as are occasional replacements of original titaniferous magnetite grains by gray translucent
masses of leucoxene. Characteristically, there is a notable scarcity of
clinozoisite or epidote in the wallrocks. Sulfide minerals, mainly finely
crystalline pyrite and rarely chalcopyrite and arsenopyrite, locally
occur in the wallrocks to a minor extent. Silicification is negligible,
but there are quartz veinlets in some specimens.

DISTRIBUTION AND RELATION TO REGIONAL GEOLOGY

In California most of the lode gold deposits in the Klamath
Mountains (pl. 2) occur in an arcuate belt approximately 130 miles
long in the east-central part of the province; the belt extends from
the California-Oregon line to the latitude of Redding. The width of
the belt ranges from about 20 to 40 miles. This gross distribution of
deposits reflects the predominant structural trends of the east-central
Klamath Mountains; some districts, however trend across the
prevalent structural direction.

Rocks in nearly all the lithologic belts are hosts for the ore bodies
(pl. 2), but the deposits are distributed unevenly among these belts.
Very few deposits have been reported in rocks of the western Jurassic
belt. The area south of Yreka that contains rocks of the eastern
Klamath belt is essentially devoid of gold lodes, and few occurrences
are known in the northeastern part of the block of the eastern Klamath belt north of Redding. The largest and most productive districts
are, however, in the southwestern part of the eastern Klamath belt
west and northwest of Redding. Lode deposits are not present in the
area of amphibolite-grade metamorphic rocks in the northern part of
the western Paleozoic and Triassic belt or in the highly foliated low-
grade metamorphic terrane on the California-Oregon border. Many
deposits, some with notable productions, lie in the eastern part of the
western Paleozoic and Triassic belt from the State boundary as far
south as lat 41°; several districts tend to cluster near the contact of
this belt with the central metamorphic belt. Numerous lodes are
known in the southern part of the western Paleozoic and Triassic
belt. North and south of Weaverville, rocks of the central metamorphic belt are hosts for deposits in several small districts. Some small
and a few moderately productive deposits are known in the large
area dominated by ultramafic, mafic, and granitic rocks in the eastern
part of the Klamath Mountains.

In Oregon the areas of gold mineralization are not as clearly
defined or restricted to apparent structural environments as in the
California part of the Klamath Mountains province. Lode deposits
are scattered with no obvious pattern throughout a wide area occupied

by rocks of the western Paleozoic and Triassic belt north of the California-Oregon State boundary. This area in itself, however, constitutes a mineralized belt which is bounded on the east by volcanic rocks of the Cascade province and on the west by rocks of the western Jurassic belt. The western Jurassic belt in Oregon is generally lacking in gold lodes except in the Greenback and Galice–Mount Reuben districts, on the east and west borders of the belt, respectively, and in the Chetco and Illinois River districts, between two northward projections of the Josephine peridotite pluton (Wells and others, 1949, p. 9). The gold mineralization in the western Jurassic belt is mostly in metavolcanic rocks. Deposits in sedimentary rocks of the Galice Formation are uncommon.

No lode gold deposits are known in the Franciscan and Dothan Formations west of the Klamath Mountains province. The small Mule Creek area in Oregon west of the province boundary is in volcanic rocks of uncertain age which are in fault contact with the Dothan Formation. These volcanic rocks are possibly correlatives of volcanic rocks in the western Jurassic belt.

It is difficult to determine the extent to which regional geology has controlled localization of the gold deposits in the Klamath Mountains. When the distribution of mineralized areas is compared with the geologic map of the province, it is apparent that in parts of the region there is some correlation with regional structural and lithologic features (pl. 2). In other places there is no apparent relation between mineralized areas and the regional geology as it is presently known.

In the southeastern part of the province the deposits of the French Gulch–Deadwood district are alined along an east-west fracture system near the crest of a broad eastward-trending arch (Albers, 1961, 1965) in rocks of the eastern Klamath belt, the Copley and Bragdon Formations. The abundance of hypabyssal minor intrusives of birdseye porphyry and quartz porphyry and the numerous gold-quartz veins led Albers (1961, 1965) to infer that an eastward-trending salient of the Shasta Bally batholith underlies the fracture zone and the archlike structure. Several of the mines in the French Gulch–Deadwood and Whiskeytown districts are along a thrust fault contact between underlying metavolcanic rocks of the Copley and Balaklala Formations and slates of the Bragdon Formation on the upper plate. Steeply dipping gold-bearing veins cut both the Copley and the Bragdon Formations but were richer and more productive where they penetrated the Bragdon Formation slates. In the Minersville and Dog Creek districts several less important mines and prospects are located near windows of Copley Greenstone exposed in slate of the Bragdon

Formation. Wherever the contact between Bragdon and Copley in these windows have been seen, it is faulted or occupied by dike rocks. The association of gold mineralization with these windows seems more than fortuitous, and it is suggested here that the contact was an avenue for the movement of mineralizing solutions.

Reconnaissance mapping by W. P. Irwin (written commun., 1968) in the southern part of the province suggests that some control has been exerted on the location of lode deposits in the Hayfork and Harrison Gulch areas by the northwest-trending contact between a major unit of metavolcanic rocks and a unit composed of siliceous metasedimentary rocks. The deposits are east of the contact, in the metasedimentary rocks.

South and east of Sawyers Bar in the Liberty district, many mines and prospects are concentrated in a small area adjacent to a contact that has been interpreted as a thrust fault between intensely deformed siliceous schist on the upper plate and underlying sedimentary and metavolcanic rocks of the western Paleozoic and Triassic belt. Most of the deposits are in metasedimentary rocks of the lower plate near the contact and are associated with minor intrusive dike rocks. Some are in metavolcanic rocks farther from the contact. A few veins are in the siliceous schists, again, however, near the contact. The northeastward continuation of this contact is found in the vicinity of Yreka and Fort Jones, where it is paralleled by a belt of gold mineralization in which, however, the deposits are farther from the contact, are more scattered and are mostly in metavolcanic rocks. The fault bounding the siliceous schists marks the south limit of mineralization, and no deposits occur in the highly deformed rocks. It is tempting to speculate on the significance of the apparent spatial relationship of the mineralized areas with respect to this contact. The fissures which localized the gold-quartz veins may have been faults that opened and reopened in response to tectonic activity responsible for deformation and overthrusting of the plate of siliceous schists. Although the main episode of thrusting is thought to have preceded the emplacement of granitic plutons, to which the gold deposits are probably genetically related, renewed or continued movement during the plutonic episode would have been responsible for opening of fractures in the lower plate at the appropriate time.

There is some concentration of gold deposits on the north and south borders of the Trinity ultramafic pluton near its contacts with rocks of the eastern Klamath belt. Of unknown significance is the fact that several deposits on the south border, in the old Carrville or Trinity Center district, are gold-sulfide replacement bodies, and on the north in the Callahan district the gold-quartz veins are rich in pyrite. The sulfide replacement deposit at the King Solomon mine west of Cecilville is in a klippe of eastern Klamath rocks above serpentinite, possibly correlative with the Trinity ultramafic body.

Concerning southwestern Oregon, Diller (1914, p. 23, 39) noted the scattered distribution of many small mines and prospects and concluded that tectonic movement previous to mineralization of the region caused widespread fracturing of the rocks instead of localizing a few fissures in narrow belts, with the result that when mineralization occurred it was widely dispersed in small veins instead of being concentrated in definite belts of lodes. His concept may have some validity for the area occupied by rocks of the western Paleozoic and Triassic belt where many small deposits are widely scattered in a shotgun pattern. Exceptions to this generalization are found, however, in the more important Greenback and Galice districts.

The small concentration of deposits in the Greenback district north of Grants Pass is on the boundary between rocks of the western Paleozoic and Triassic and western Jurassic belts. The boundary here is marked by a complex pattern of serpentinite and gabbroic rocks which are believed to have been intruded along a thrust fault zone. Granitic plutons southwest and northeast of the Greenback district were also possibly emplaced along this zone. Other fissures in this zone may have guided mineralizing solutions genetically related to the granitic intrusives.

The Galice district on the northwest boundary of the Klamath Mountains province is in an area where there has been extensive faulting and emplacement of ultramafic, gabbroic, and granitic rocks. Mineralization has been along fractures that parallel the major northeast-trending faults; the veins penetrate all the rocks, including the intrusives, on the southeast side of an eastward-dipping thrust fault that constitutes the boundary between rocks of the western Jurassic belt and the Dothan Formation. Most of the deposits are in metavolcanic rocks of the Rogue Formation and in amphibolite that Wells and Walker (1953) regarded as the metamorphic equivalent of the Rogue. Few occur in sedimentary rocks of the Galice Formation that lie to the east of the Rogue Formation. Gold-bearing veins are not found in the Dothan Formation, and the thrust fault is tentatively regarded as being younger than the episode of gold mineralization.

The Mule Creek district in northeastern Curry County and southeastern Coos County, Oreg., is an isolated area west of the main Klamath Mountains belt of gold mineralization. Deposits of the Mule Creek district are in metavolcanic rocks that have tentatively been correlated with the Rogue Formation of the western Jurassic belt

(Wells and Peck, 1961). The metavolcanic rocks are intruded by gabbro and diabase, and their east contact with the Dothan Formation is a fault; to the west they are overlapped by uppermost Jurassic, Lower Cretaceous, and Tertiary sedimentary rocks. The absence of mineralization in rocks of the Dothan Formation between the Mule Creek district and the western Jurassic belt is evidence that the sedimentary rocks of the Dothan Formation are younger than the episode of mineralization that affected rocks elsewhere in the Klamath Mountains.

Granitic plutons are widespread in the Klamath Mountains province, but there is no regular, obvious, spatial or size relationship between the distribution of individual lodes or districts and exposures of granitic intrusions. Granitic plutons are as plentiful and as large, apparently, in the western part of the province where gold deposits are scanty as they are in the central and eastern parts where gold mineralization is more common.

A large area in the north-central part of the Klamath Mountains province is essentially devoid of gold-bearing lodes. In this region the subcircular area occupied by the schists of Condrey Mountain is surrounded by large irregularly shaped areas underlain by amphibolites and siliceous schists accompanied by many intrusive bodies of serpentinite. The region is not totally unmineralized, for there are several small quicksilver deposits in the schists of Condrey Mountain, especially near the fault contact with the surrounding more highly metamorphosed rocks, where a few small gold lodes have also been discovered. A similar area of siliceous schist and amphibolite in Oregon near the northeast border of the province is also lacking in gold lodes except at its south end. On the northwest boundary of the province, however, similar rocks of the same metamorphic grade in the Galice area are abundantly mineralized. The reasons for lack of mineralization in these areas are obscure.

AGE OF MINERALIZATION

The gold-bearing lodes of the Klamath Mountains occur in all the main subjacent rock units and in many of the plutonic igneous rocks; some are closely associated with dikes and sills that are genetically related to the intrusives. There is no evidence to indicate a difference in age between the gold-quartz veins and the gold-sulfide replacement bodies. It is probably safe to assume that gold-bearing veins and replacement bodies are related in origin to the granitic plutons. The fractures which guided the flow of the mineralizing fluids and provided openings in which the vein-forming materials were deposited probably formed during the Late Jurassic Nevadan orogeny. No primary deposits are known in the superjacent rocks.

LODE GOLD DISTRICTS IN CALIFORNIA
DEPOSITS IN THE VICINITY OF REDDING

The greatest concentration of lode gold deposits in the Klamath Mountains is in the southern part of the province near Redding, Calif. (pl. 2). The geology of much of this area has been thoroughly studied by Kinkel, Hall, and Albers (1956) and Albers (1964, 1965), from whose work the following information has been taken. Earlier geological studies were made by Diller (1906) and Ferguson (1914), and Averill (1933, 1939) reported on the mines.

Many of the mines are in sedimentary and volcanic rocks of Paleozoic age, and a substantial number are in granitic rocks that intrude the Paleozoic strata. The Paleozoic rocks are intruded by two granitic plutons, the Mule Mountain stock and the Shasta Bally batholith. The stock is mainly trondhjemite and albite granite; the batholith is quartz diorite and granodiorite. Mafic and felsic dikes and sills intrude the Mule Mountain stock as well as the Paleozoic rocks. Of these minor intrusives, andesite porphyry, diorite porphyry, and dacite porphyry are closely associated with many of the gold deposits.

FRENCH GULCH–DEADWOOD DISTRICT

French Gulch, a small town approximately 15 airline miles northwest of Redding, is midway in a mineralized zone 1 to 2 miles wide and approximately 10 miles long that trends almost due west from the Gladstone mine in Shasta County to the Venecia mine in Trinity County. The west end of the zone, in Trinity County, is known as the Deadwood district; the central and eastern parts, in Shasta County, are the French Gulch district. The combined district was one of the most productive gold areas in the Klamath Mountains. Five mines produced more than $1 million worth of gold each, with the Gladstone, at the east of the district, producing 206,765 ounces of gold (about $4 million), and the Brown Bear, near the west end, between 400,000 and 500,000 ounces ($8 and $10 million). At least seven other mines produced $100,000 to $500,000 each (5,000 to 25,000 oz) in gold.

Albers (1961; 1965, p. 19) described the mineralized belt as occurring along an east-west fracture zone in the Bragdon Formation on the south flank of a crudely defined eastward-trending arch. The rocks are cut by numerous faults and intruded by sills and dikes. Irregular intrusives of birdseye porphyry are abundant along the mineralized zone (Albers, 1964, 1965). At some places sills of birdseye porphyry and quartz porphyry occupy the contact between the Copley and overlying Bragdon Formations. Discontinuous, irregularly shaped exposures of Copley Greenstone underlying the Bragdon Formation occur along the fracture zone. Albers' interpretation (1964, p. J63, pl. 1) is that these areas of Copley Greenstone are exposed in windows eroded in a thrust plate of Bragdon Formation. The thrust fault is exposed in part but in many places has been modified by later normal faulting and occupied by intrusive bodies of porphyry.

The gold deposits are of two main types (Albers, 1961, p. C3) : (1) moderate to gently dipping quartz and quartz-calcite veins along the thrust contact below the Bragdon Formation, and (2) steeply dipping quartz and quartz-calcite veins in the Bragdon Formation. Birdseye porphyry intrusives are commonly associated with both types and are cut by the veins. Most of the gold from the Washington mine and many small pocket deposits came from veins of the first type. The most productive mines in the district, including the Brown Bear, Niagara, American, Gladstone, and Summit and Montezuma, were developed on veins of the second type. These veins were richest at intersections with veins of opposite dips and at intersections of veins with contacts between Bragdon shale and porphyry. Apparently the Bragdon Formation was more favorable for the localization of economic deposits, because where steep veins cut the Copley Greenstone, they are commonly narrow and of poor grade, while in the Bragdon and at its contact with the Copley, they are productive. Because of the possible significance of carbon or graphite as a precipitating agent for gold, as suggested by Hershey (1910), and Ferguson (1914, p. 40–43), the composition of the shales of the Bragdon Formation may have been partly responsible for veins in that formation being more economic than in the Copley.

Gold occurs mainly free in the veins but is also in the sulfide minerals, which include pyrite, galena, sphalerite, arsenopyrite, and rarely, chalcopyrite. The chief gangue is quartz, but minor amounts of calcite and mica are also present (Albers, 1961, p. C3).

WHISKEYTOWN DISTRICT

The Whiskeytown district is approximately 7 miles southeast of the French Gulch district and extends in a general northwest direction north of Whiskeytown Lake. (The old settlement of Whiskeytown is now below the surface of the reservoir.) One mine in the district, the Mad Mule, produced in excess of $1 million worth of gold (about 50,000 oz) ; three others, the Gambrinus, Mount Shasta, and Truscott mines, are known to have produced on the order of from

$60,000 to $175,000 each, and there were many other mines and prospects whose production is not known but is believed to have been relatively small.

The northwestern part of the district is underlain by the Copley Greenstone and Balaklala Rhyolite, which are in contact with the Bragdon Formation along the Spring Creek thrust fault. Rocks of the Bragdon Formation form the upper plate of the thrust. Altered granitic rocks of the Mule Mountain stock, which intrudes the Copley and Balaklala, constitute the southeast end of the district.

A few deposits occur along the Spring Creek fault, including that of the Mad Mule mine, which had the largest production in the district. Some along this contact, like those in the French Gulch–Deadwood district, are on contacts between birdseye porphyry and shales of the Bragdon Formation. Ferguson (1914, p. 40–43) classified two deposits along this contact, those at the Mad Mule and Eldorado mines, as pocket deposits which contained local rich concentrations of gold which he believed formed by secondary enrichment. Numerous prospects occur between the Spring Creek thrust and the Mule Creek stock, an area underlain by Copley Greenstone and Balaklala Rhyolite, but little is known about them. Several mines and prospects are in the Balaklala Rhyolite and Copley Greenstone on or close to the west contact of the Mule Mountain stock, and some prospects are within the stock itself. Most important were the Mount Shasta and Gambrinus mines, which were on steeply dipping northwest-striking quartz veins in Balaklala Rhyolite. The veins contained free gold, pyrite, and, locally, chalcopyrite; Ferguson (1914) reported minor amounts of molybdenite at the Mount Shasta mine. According to Albers (1965, p. 32) the small deposits within the stock are in east-striking veins.

SHASTA-REDDING DISTRICT

Between Redding and the old mining town of Shasta (now a State historical monument), 5 miles to the west, and southwest of Redding as far as Centerville, were many small mines and prospects. Production from these deposits is essentially unknown, but apparently none produced more than a few thousand dollars worth of gold.

Bedrock in this area is Copley Greenstone which is intruded by the Mule Mountain stock. A recent geologic map of the Redding quadrangle (Hollister and Evans, 1965) shows lenticular bodies of argillite within the greenstone and some lenticular to tabular bodies of felsic volcanic rock which have been mapped as Balaklala Rhyolite. Numerous small dikes of birdseye porphyry intrude the Copley and

Balaklala in a zone as much as 1 to 1½ miles wide adjacent to the east contact of the stock, but not the stock itself.

The deposits on which the prospects and small mines were located are in the Copley Greenstone, in a belt adjacent to the Mule Mountain stock. Several deposits also occurred within the stock near its east contact. The gold occurred in quartz veins, many of which were accompanied by birdseye porphyry dikes, particularly in the Bragdon Formation. From a limited amount of data, it appears that many of the veins in the Copley Greenstone strike northeast to north and have vertical to steep easterly dips, thus paralleling the average trend of planar structures in the country rock. In the stock the veins have west to northwest strikes.

MULETOWN AND SOUTH FORK DISTRICTS

The Muletown and South Fork districts are two small areas of mineralization 2 and 3 miles north and northwest, respectively, of Igo, a small settlement 10 miles southwest of Redding. The two districts are of interest mainly because they have contrasting characteristics of mineralization. Deposits of the Mule Mountain district are gold bearing, while those of the South Fork district have mainly yielded silver and minor quantities of gold. Neither district has produced much gold, but the Silver Falls–Chicago Consolidated mine in the South Fork district is notable for having produced more than $1 million in silver.

The Muletown district is at the south end of the Mule Mountain stock. The country rock intruded by the Mule Mountain pluton is the Copley Greenstone, which is also cut by dikes of Balaklala Rhyolite. A small body of granodiorite which Albers (1964, p. J37, pl. 1) called the Clear Creek plug intrudes the Copley and the Mule Mountain stock in this region. The gold-bearing veins have northeasterly strikes and occur in both the Copley Greenstone and granitic rocks of the Mule Mountain stock and the Clear Creek plug. Many are close to contacts of the intrusion, occurring in either the Copley or granitic rock. The veins were quartz and contained free gold and pyrite.

The South Fork district is from 1 to 4 miles west of deposits of the Muletown district and almost entirely within the Shasta Bally batholith close to its southeast contact with the Copley Greenstone. Most of the veins strike northeast and dip steeply, mainly southeast. The principal minerals include tetrahedrite, galena, sphalerite, pyrite, native silver, gold, and chalcopyrite (Tucker, 1922, p. 315). Gold has been of minor importance compared with silver in this district. However, the Great Falls mine (Ballou or Manzanita) is unique in that it produced only gold in a mainly silver producing district.

OLD DIGGINGS AND ADJACENT DISTRICTS

The Old Diggings district is on the east side of the Sacramento River approximately 5 miles north of Redding. Originally an active placer-mining district, it later achieved some value as a lode district because the nearby copper smelters demanded large quantities of silica for use as a flux in their smelting process. Some high-grade gold ore was mined and milled, but much of it was relatively low grade and not free milling; however, the demand for quartz at the smelters made profitable the mining and shipping of otherwise non-economic material. Gold production from the district is poorly known, but one mine, the Reid, is reported to have produced about $2.5 million (Logan, 1926, p. 178). Other major producers were the National ($200,000), Central ($500,000), and Texas Consolidated ($750,000) mines. Two other districts of minor economic importance are included here with the Old Diggings district because the gold deposits have a similar mode of occurrence. These are the Quartz Hill district, about 2 miles south of the central part of the Old Diggings district, and the Flat Creek district, west of the Sacramento River north of the north end of the Mule Mountain stock.

The Old Diggings district is underlain by the Copley Greenstone and several thin tabular north-northeast-trending masses of Balaklala Rhyolite (Kinkel and others, 1956, pl. 1). In the Quartz Hill and Flat Creek districts there are fewer bodies of Balaklala Rhyolite, and in the Flat Creek district they strike northeast. The country rock is cut by numerous parallel quartz veins that strike northeast and have steep southeastward to essentially vertical dips in the Old Diggings district. In the Quartz Hill district some of the veins strike northwest, while in the Flat Creek district they strike nearly west. The veins contained some free gold but also considerable pyrite and chalcopyrite with which the gold was associated.

DOG CREEK (DELTA) DISTRICT

The Dog Creek district is approximately 25 miles north of Redding in the headwaters area of Dog Creek, a tributary of the Sacramento River, and 2½-5 miles west of Delta, a station on the Southern Pacific railroad. To the southwest, some prospects on Stacey Creek, southwest of Tollhouse, are included in the Dog Creek district. Production from the district has been small, amounting to a few tens of thousands of dollars at most. The most important mine was the Delta Consolidated with a recorded production of $32,000 (Ferguson, 1914, p. 71). Most of the properties seem to have been little more than prospects; however, a few had small high-grade pockets of free gold.

The Dog Creek district is of interest mainly because it is in Copley Greenstone and Balaklala Rhyolite exposed through windows in the Bragdon Formation. Therefore it may be structurally similar to the French Gulch–Deadwood district.

According to descriptions by Ferguson (1914, p. 71–73) and Brown (1916, p. 777–804), most of the gold-bearing quartz veins strike nearly west and are vertical or dip steeply north or south. They cut the Copley, and some penetrate the overlying shales of the Bragdon Formation. Quartz porphyry dikes are prevalent in the district and are reported to accompany some of the veins (Ferguson, 1914; Brown, 1916). Besides free gold, the veins commonly contain pyrite, and chalcopyrite is locally abundant; small amounts of galena and sphalerite have also been reported.

Three samples collected from the contact between slate and greenstone in the Dog Creek district contained no gold, but two contained traces of silver (table 1).

BULLY CHOOP (INDIAN CREEK) DISTRICT

Two small clusters of deposits, the Indian Creek and Bully Choop district, located about 10 and 15 miles southeast of Weaverville, respectively, are collectively referred to herein as the Bully Choop district (pl. 2). Production is essentially unknown, but probably amounted to no more than a few tens of thousands of dollars.

The district is west of the large granitic Shasta Bally batholith in rocks of the central metamorphic belt, which here include both the Salmon Hornblende Schist and the Abrams Mica Schist (Irwin, 1963). It may be significant that both the Indian Creek and the Bully Choop groups of deposits are on the crests of two northwest-trending antiforms (Irwin, 1963).

The gold-bearing lodes were narrow quartz veins, generally ranging from thin stringers to veins a few inches to a foot — seldom as much as several feet — in width, mostly with northwest strikes and steep dips. Some of the gold was free milling, but mostly it was contained in sulfide minerals, chiefly pyrite but including minor quantities of galena, chalcopyrite, and sphalerite. Some was mined from small high-grade pockets near the surface, where weathering had caused decomposition of gold-bearing sulfides.

HARRISON GULCH DISTRICT

The Harrison Gulch district is mostly in southwestern Shasta County near the Trinity-Shasta County line (pl. 2). It centers around the now-abandoned Midas mine, which is approximately 2 miles north of State Route 36 and approximately 4 miles east of the settlement of Wildwood. State Route 36 links Interstate 5 near Red Bluff with U.S. 10 near Eureka.

The district is not a large one, embracing only a few abandoned prospects and one mine, the Midas, which, at one time, was one of the most valuable properties in Shasta County and the Klamath Mountains province. A production of more than $3.5 million worth of gold was reported for the Midas mine between 1896 and 1914 (Logan, 1926, p. 173). Total production was more than $4 million (Averill, 1939, p. 142).

Little is known about the geology of the district other than that the district is underlain by rocks of the western Paleozoic and Triassic belt. The beds strike northwest, and they dip moderately to steeply northeast. Northwest of the district a granitic body of uncertain size and shape intrudes the metavolcanic and sedimentary strata. Very small intrusive bodies of granitic rock and serpentinite are exposed in a roadcut near the remains of the old millsite at Midas mine.

The Midas mine was developed on three parallel veins with northwest strikes and moderate to deep southwest dips; these veins are now marked by alined adits and shafts on the northwest side of Harrison Gulch. The country rock is chert and slaty argillite; diabase is also reported to occur in the vicinity of the veins (Logan, 1926, p. 174). Apparently the gold was in quartz veins which occurred as lenses in the sheared country rock (Logan, 1926, p. 174).

Little is known concerning the other properties in the district except that they were located on small quartz veins in "slate," some of which had diabase or quartz porphyry on one wall.

HAYFORK DISTRICT

The Hayfork district includes a small isolated group of deposits 3–5 miles southeast of the town of Hayfork, Trinity County, (pl. 2). The mines are small, and production was probably low. One of the properties, the Kelly mine, was in production in 1966. Presumably the very rich placers of the Hayfork River derived their gold from this area.

Mines and prospects of the Hayfork district are in rocks of the western Paleozoic and Triassic belt which here include argillite, siliceous shale and chert, some graywacke, minor greenstone, and occasional lenticular bodies of limestone or marble. According to W. P. Irwin (oral commun., 1968), the mines and prospects in this area are in the sedimentary rocks near a contact with a belt of metavolcanic rocks that adjoins the sedimentary rocks on the west. The deposits consisted of small quartz veins containing free gold and some auriferous pyrite.

The principal mines were the Mueller (also referred to as Layman or Hayfork) and Kelly mines. Little is known about the Mueller, which is approximately 1½ miles southeast of the Kelly, except that for many years it was a small producer. The Kelly mine, on a deposit discovered about 1932, has been operative to the present except for an idle period from 1959 to 1961. Underground workings on four levels are not extensive, but the mine has had a substantial production by selective mining of high-grade ore. The gold occurs free, much of it as handsome specimen rock, in white quartz. Quartz veins and lenses that range from approximately 1 inch to as much as 12 inches in width in black sheared argillite, strike N. 25° to 30° W., dip eastward at moderate to gentle angles, and are intersected and displaced by northeasterly striking, moderate to steep eastward-dipping postore faults. In addition to gold, the quartz contains some pyrite and galena. According to the present owner, a northwest-southeast-trending zone of mineralization extends 1½ miles to the Mueller mine.

MINERSVILLE DISTRICT

The Minersville district includes the country along the Trinity River and Stuart Fork from the vicinity of Trinity Dam northward for approximately 10 miles (pl. 2). The settlement of Minersville, for which the district was named, approximately 10 miles northeast of Weaverville, and many of the mines and prospects are now covered by the water of Clair Engle (Trinity) Lake. Most of the deposits were exploited by small mines and prospects whose production is unknown; however, two properties, both now under water, the Fairview and Five Pines, had a fair production.

Geologic maps of the area made prior to flooding by Clair Engle (Trinity) Lake (Ferguson, 1914, pl. 2; Hinds, 1933, pl. 3; Irwin, 1960b, pl. 1) show that the Minersville district is underlain by irregularly shaped areas of Copley Greenstone surounded by the overlying Bragdon Formation. In this respect, the district is similar to the French Gulch–Deadwood district to the southeast. Ferguson (1914, p. 73) attributed exposures of the Copley Formation to erosional breaching of irregular folds by the Trinity River and its tributaries. Possibly the contact between the Bragdon and Copley Formations is a thrust fault, as Albers (1964, p. J62–J64; 1965, p. 16) postulated for the French Gulch-Deadwood district, and the areas of Copley Greenstone are fensters in a thrust plate. In contrast to the French Gulch–Deadwood district, minor intrusive bodies such as birdseye porphyry are scarce.

Most of the gold-bearing lodes were on the contact between slate and greenstone or in slate near the contact. None of the veins seems to have been persistent horizontally or vertically. Most of the values were obtained from erratically distributed high-grade pockets. Ferguson (1914, p. 41–42) observed that many deposits were typical pocket deposits which occurred close to the surface and were almost exclusively along greenstone-slate contacts.

Samples were collected from roadside exposures at several places in the Minersville district southwest of Clair Engle (Trinity) Lake (fig. 5). Analyses (table 2) show that some contain small amounts of gold and anomalously high amounts of some other metals. Samples from two porphyry dikes east of Clair Engle (Trinity) Lake near the Five Pines mine contained trace amounts of gold.

TRINITY CENTER DISTRICT

The Trinity Center district, as used here, includes the old Trinity Center, Carrville, Coffee Creek, and Dorleska districts, in northeastern Trinity County near the town of Trinity Center (pl. 2). The mineralized area occupies a poorly defined belt approximately 20 miles long, extending in a west-northwest direction from the East Fork of the Trinity River to the headwaters of Coffee Creek, a tributary of Trinity River. Many small mines and prospects were located in the district, but few produced more than a few thousand dollars worth of gold, except the Bonanza King mine, at the southeast end of the mineralized belt, which is reported to have produced $1.25 million worth of gold.

The geology of the district is known mostly from reconnaissance studies by geologists of the Southern Pacific Company. In recent years the western part of the mineralized belt was mapped by Davis, Holdaway, Lipman, and Romey (1965). Several of the mines and prospects were described in articles by Hershey (1899, 1900) and MacDonald (1913).

The Trinity Center district is in the southern part of a large complex composed predominantly of peridotite and serpentinized peridotite in which there are numerous irregularly shaped plutons that range in composition from gabbro to quartz diorite and granodiorite. The southeastern part of the district is near a northwest-projecting reentrant of eastern Klamath Paleozoic rocks which is bounded by northwest-striking faults. Approximately 11 miles west

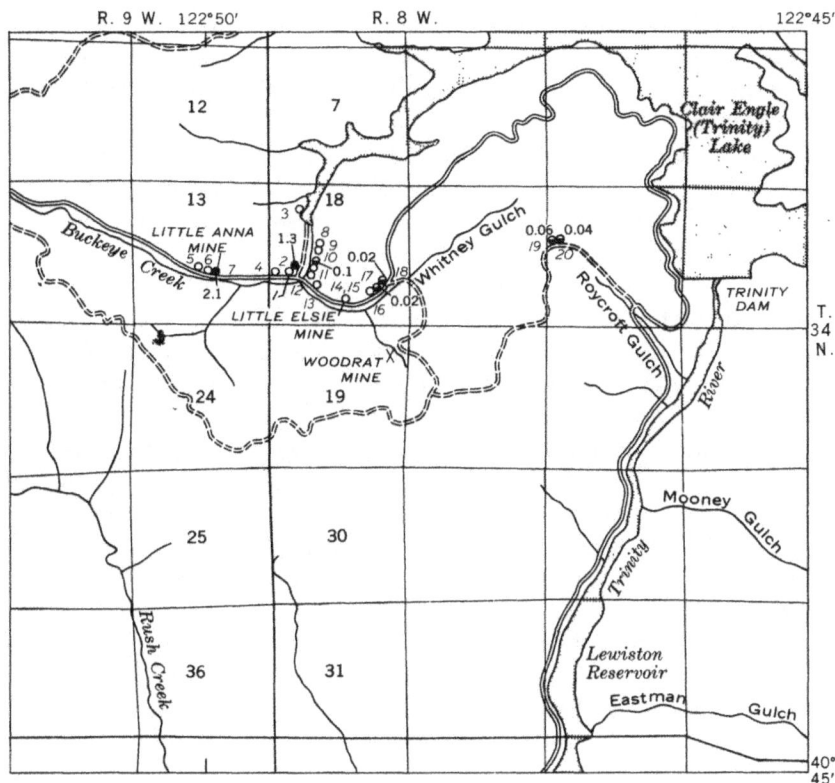

FIGURE 5.—Sample localities, Minersville district, California.

of Trinity Center, deposits of the old Dorleska district are near the intersection of an east-west fault and the boundary between the ultramafic complex and rocks of the central metamorphic belt.

The varied nature of the country rocks and the complexities of their distribution are reflected in the diverse modes of occurrence of deposits in this mineralized district. MacDonald (1913) classified the deposits into five types, but probably there are only two main kinds, replacement deposits and those that occur as veins. The vein-type deposits can be further subdivided according to the country rock in which they occur. Common to all sites of mineralization, however, are basaltic and lamprophyric dikes which, besides intruding older rocks in the vicinity of an ore deposit, are frequently mineralized themselves or along their contacts. Dikes of dacite porphyry, commonly referred to as birdseye porphyry, are also common.

Approximately 1¼ miles southeast of the old settlement of Carrville, near the old Headlight mine, rocks of the eastern Klamath belt project into the ultramafic complex. A common feature of several old mines and prospects in this part of the district (MacDonald, 1913) is the occurrence of ore as replacements of mafic dikes where they intrude irregularly shaped bodies of dacite porphyry. At some places the mafic dikes intrude serpentinite. At several places the country rock into which the dikes are intrusive is greenstone, probably the Copley Greenstone, which is in contact with slates of the Bragdon Formation. The mineralized zone is sheared, and though the primary ore seems to have been low grade, weathering has enriched its surficial parts. At the Headlight mine the ore zone was gently dipping, and there is some suggestion that, as in the French Gulch–Deadwood district (Albers, 1965), mineralization occurred on or close to the contact between slate of the Bragdon Formation and the Copley Greenstone.

Several deposits in the eastern part of the Coffee Creek valley, represented by that of the Golden Jubilee mine, occur as gold-bearing quartz veins in granitic rocks. They are commonly accompanied by lamprophyre dikes which in some places constitute one of the vein walls. The prevailing trend of the veins and the dikes of lamprophyre is northeast, about at right angles to the apparent general trend of the Trinity Center mineralized area.

At the west end of the area, the Dorleska and two other mines were developed on quartz veins in serpentinite. Again, accompanying the gold-bearing veins are lamprophyre dikes and a wide dikelike body of dacite porphyry which also intrude the ultramafic rock. The veins are said to have been along shear zones in and along the contacts of the lamprophyres (MacDonald, 1913). They trended northeastward. Gold occurred in quartz and in the shear zones without quartz.

The Bonanza King mine, at the southeast end of the mineralized belt, is high on the south slope of Bonanza King Mountain in the southern part of a large gabbroic pluton in ultramafic rocks. Little is known about the occurrences here aside from observations made during the recent reconnaissance. The old workings are in gabbro which is widely variable in texture and composition and locally contains large blocks and small inclusions of serpentinized peridotite. MacDonald (1913), though he did not visit the property, reported that the country rock was greenstone cut by lamprophyric dikes. Neither rock type was seen during the recent examination. Material on the dumps is gabbro containing minor sulfide and some quartz. MacDonald (1913) reported that the lode was a mineralized shear zone in which were localized high-grade ore shoots.

CANYON CREEK–EAST FORK DISTRICT

The Dedrick–Canyon Creek and Helena–East Fork districts, in Trinity County, constitute a single mining district whose center is approximately 12 miles northwest of Weaverville (pl. 2). For convenience, the combined districts are here referred to as the Canyon Creek–East Fork district.

Production from lode mines of the Canyon Creek–East Fork district has been modest. The East Fork mines have achieved the greatest output; three of them, the Enterprise, Golden Crest, and North Star, have recorded productions of more than $100,000 each; the Alaska mine produced more than $600,000. The Globe mine, on the slope and ridge east of Canyon Creek, is estimated to have produced more than $700,000. In addition there are several small mines and prospects whose production is unknown or less than $100,000.

Mines and prospects of the Canyon Creek–East Fork district are in rocks of the central metamorphic belt, here composed exclusively of the Salmon Hornblende Schist. A granitic body of moderate size, the Canyon Creek pluton (Davis and others, 1965), intrudes the hornblende schist north of the Canyon Creek district, and other smaller granitic bodies occur in the southeastern and southwestern parts of the district. The geology of the district is not well known, and there are no apparent geologic reasons for localization of mineralization in this area. Little could be learned from the brief field examination of the region. Several of the old mines were difficult to locate, and some could not be found. Apparently all were near quartz veins that contained free gold accompanied by pyrite, some of which at least was gold bearing. Felsic dikes, including quartz porphyry and granitic rocks are associated with several of the deposits. Analyses of samples cut from quartz veins near the Globe mine are listed in table 3.

NEW RIVER–DENNY DISTRICT

The New River–Denny district is in northwestern Trinity County at the crest of the Salmon Mountains divide between the drainages of the Trinity and Salmon Rivers, near the Trinity-Siskiyou County line (pl. 2). The area is in exceedingly rugged terrain at the headwaters of New River, a tributary of the Trinity River. It is in the Salmon–Trinity Alps Wilderness Area, accessible only by trail. Production from the area has been relatively small, probably $1 million or less. Recorded gold production is valued at about $750,000.

The geology of the area is known mainly from reconnaissance by Irwin (1960b) and from more detailed mapping in areas adjoining it to the east (Davis, 1968) and southeast (Cox, 1956). The area is in rocks of the western Paleozoic and Triassic belt approximately 1–3 miles west of a long, narrow serpentinite body which Davis (1968, p. 918) believes occupies a low-angle eastward-dipping thrust fault which he named the North Fork fault zone. The country rocks are north-trending mafic volcanic rocks and diabase. Farther west, mostly outside the mineralized area, the rocks are chert, argillite, quartzite, and bodies of lenticular limestone.

Most of the gold was produced from narrow discontinuous quartz veins and shear zones partly occupied by quartz. Many of the veins and shear zones are at the contact between diabase and quartz porphyry dikes; some have diabase on both walls of the fracture, and a few are reported to be within quartz porphyry dikes. Most of the veins strike northwest and dip moderately to steeply northeast. The gold occurred free in the quartz; the bulk of the ore was low grade, and small high-grade shoots accounted for most of the production. In addition to gold, pyrite and some chalcopyrite and arsenopyrite occurred in some mines.

LIBERTY (BLACK BEAR) DISTRICT

The Liberty (Black Bear) district is in southern Siskiyou County near Sawyers Bar, a village on the North Fork of Salmon River (pl. 2). The majority of the deposits as well as the most productive ones occupy an arcuate belt approximately 12 miles long and ½ to 1 mile wide on the steep slopes south and east of the river. Several prospects and small mines occur on the slopes of Tanners Peak north of the river and northeast of Sawyers Bar. One of the most productive mines in the Klamath Mountains, the Black Bear, with a recorded output of more than $3 million in gold, is at the west end of the arcuate belt of mineralization. In addition, the belt includes several mines that produced from $100,000 to more than $500,000 in gold.

South and east of Sawyers Bar, rocks of the western Paleozoic and Triassic belt are overlain by highly deformed siliceous schists which occupy the divide between the North and South Forks of the

Salmon River. The rocks north and west of the schists are interbedded greenstones, argillites and slaty argillites, cherts, and occasional small lenticular bodies of recrystallized limestone. Many of the greenstones are massive and virtually structureless, but at several places relics of original pillows and volcanic breccias are preserved. Often the cherts are rhythmically bedded with siliceous layers 1 or 2 inches thick. The argillaceous rocks close to the schist belt generally are phyllitic, and the rhythmically bedded cherts are tightly folded. The siliceous schists are contorted quartz-muscovite schists and phyllites and micaceous quartzites. The contact between the schists and underlying undeformed rocks is not known, but the contact is probably a thrust fault whose inclination ranges from gentle to steep southward. The belt in which most of the mines are concentrated is adjacent to this hypothetical fault.

Most of the deposits in the arcuate belt south and east of Sawyers Bar are in rocks of the western Paleozoic and Triassic belt below the contact with the siliceous schists; many are near the contact, and some may be on it. A few prospects and small mines are in the siliceous schists above the contact. Thus it appears that rocks of the western Paleozoic and Triassic terrane were more favorable for the formation of auriferous veins than the siliceous schists. There seems to be no preferred lithology for location of the veins, for they occur in greenstone and argillaceous rocks. It may be significant, however, that the Black Bear mine, the most productive in the district, as well as one of the most productive in the Klamath Mountains, was in argillite. It was also noted during field examination that dark-gray to black usually sheared argillite is a common rock type on many of the old mine dumps, and the old reports mention that black carbonaceous slate constitutes the wallrock in many of the mines.

In earlier descriptions of the mines and prospects, porphyry and light-colored dikes are frequently mentioned as being associated with the gold-bearing veins. Commonly they are described as forming the hanging or footwall of the veins. A fine-grained light-colored, brown-weathering andesitic rock seen on many of the mine dumps and also as dikes intruding the country rock is possibly what has been referred to in the old reports. The dike rock is altered, composed of randomly oriented laths of albite and interstitial chlorite, irregular patchy calcite, and white translucent needlelike prisms, possibly leucoxene, which may be relicts of an original amphibole. Generally these light dikes contain small scattered grains of pyrite. In at least one place similar dikes were hosts for quartz veins which presumably were gold bearing. Granitic dikes, also containing minor amounts of pyrite, occur in some places, and in one or two localities, they are cut by quartz veins that were mined for gold. Less commonly seen are dikes of dark fine-grained unaltered lamprophyric rocks.

The deposits of the Liberty district are in fairly well defined lodes associated with quartz veins, most of which strike north to northeast and dip moderately to steeply to the east and southeast, generally normal to the trend of the contact between schist and the less deformed rocks beneath. A few trend approximately east and dip southward. Calcite in minor to plentiful amounts is usually associated with the veins, and minor amounts of sulfides, mainly pyrite, are likewise common. The gold was mostly free milling and occurred in shear zones in the country rock as well as in the quartz. Some of it came from auriferous pyrite. At the Black Bear mine, two parallel quartz veins approximately 400 feet apart were mined. The veins had northeast strikes and dipped moderately to steeply southeast. Each vein has an ore shoot approximately 400 feet long that pitches to the northeast; the shoots were partially mined to about 1,000 feet down the dip. The gold occurred free in quartz and was accompanied by very little pyrite.

The prospects north of the river on Tanners Peak were not visited, and little is known about them from the literature. They are in an area which is predominantly greenstone.

Analyses of two suites of samples from roadside exposures of fault and alteration zones are contained in table 4.

CECILVILLE DISTRICT

Lode deposits of two kinds in two different geologic environments occur in the South Fork of the Salmon River drainage west of the settlement of Cecilville (pl. 2). The first is the usual quartz-vein type marked by a rather isolated group of mines and prospects known as the Gilta or Knownothing district, approximately 10 miles northwest of Cecilville. The district had a moderate production, the Gilta and Knownothing mines having produced in excess of $500,000 and $100,000, respectively. The other type, at the King Solomon mine, 3 miles northwest of Cecilville, was a sulfide replacement body whose oxidized surficial part, or gossan, was mined by open-pit methods. Production is estimated to have been $679,000.

The vein deposits of the Gilta district are in rocks of the western Paleozoic and Triassic belt. Few details about the geology are known, but a brief reconnaissance showed that in the vicinity of the old mines argillite and slaty argillite predominate, with minor amounts of metavolcanic rock and lenticular bodies of marble. On the mine dumps the predominant rock type is sheared argillite. Some altered lamprophyric dike rock was also seen on the dump of the

Knownothing mine. On the slopes southwest of the Gilta mine a dikelike body of dioritic rock intrudes the argillite. Nothing is known about the strike and dip of the veins.

The area near the King Solomon mine has been mapped recently by Davis and Trexler (Davis, 1968, pl. 1). The deposit is in a plate of rocks composed of cherts, slates, phyllites, and sandstones of the eastern Klamath belt which has been thrust over metamorphic rocks of the central metamorphic belt and possibly rocks of the western Paleozoic and Triassic belt. Davis (1968) regards these rocks as belonging to the Duzel Formation of Ordovician (?) age.

Exposures in the mine area are poor, and in the opencut itself sloughed-in overburden has covered most of the bedrock. Averill (1935, p. 294) stated that the ore occurred in schist intruded by diorite dikes and was slightly oxidized. According to another description (Thompson, 1957, p. 14) the ore was in "a wide zone of silicified limestone ... [which contained] ... "disseminated gold values." Possibly the mineralized zone was in silicified marble or calcareous schist, because rock now exposed in the southeast wall at the south end of the opencut is a tactite composed of calcite, quartz, garnet, chlorite, and epidote with less than 1 percent of mostly oxidized pyrite. A fine-grained hornblende-biotite lamprophyre dike intrudes the tactite. Blocks of siliceous marble are exposed in the western part of the cut. A small amount of altered, epidotized calcareous mica schist also crops out there. Weathering, accompanied by oxidation and leaching of auriferous sulfides in the mineralized zone, resulted in formation of a soft, easily mined gossan.

The King Solomon ore body trended northwest and dipped steeply southwest; a northwest-striking fault dipping moderately to the southeast bounds the mineralized zone on the west. Oxidized ore of two west-pitching shoots was mined by opencut methods, leaving a pit approximately 500 feet long, 100 feet wide, and 50 feet deep. This ore contained approximately 0.1 ounce gold per ton of rock. The unoxidized sulfide ore ranged from 0.02 to 0.1 ounce gold per ton. Mining ceased in 1940 when the easily mined oxidized ore was exhausted and hard low-grade sulfide protore was reached.

Half a mile southeast of the mine, in Mathews Creek, dikelike bodies of dacite porphyry and porphyritic quartz diorite crop out. At one place near where the mine road crosses Mathews Creek the quartz diorite is altered to a greenish rock containing several percent of fine-grained pyrite in veinlets and as disseminations.

CALLAHAN DISTRICT

Several small mines and prospects, including the Dewey and Cummings (McKeen) mines, which had reported outputs of $900,000 and more than $500,000 in gold, respectively, were located in the vicinity of Callahan, a village on the Scott River at the south end of Scott Valley (pl. 2). These deposits form a poorly defined arclike zone concave to the northeast, and this arc lies along the boundary between the large complex of ultramafic, gabbroic, and granitic rocks in the eastern part of the Klamath Mountains and the south end of the large isolated plate of sedimentary rocks of the eastern Klamath belt that extends southward from Yreka. Away from this arc, a few mines occur in or near the narrow septum of rocks of the central metamorphic belt between ultramafic rocks and a large granitic pluton to the west.

Although the mines and prospects have produced mainly gold, the district is characterized by copper mineralization in several places. Typically, the veins in the Callahan district are accompanied by abundant sulfide minerals, mainly pyrite and pyrrhotite, with lesser amounts of chalcopyrite and galena. Gold reportedly occurred both free and in the sulfides. Production from several of the sulfide-bearing deposits was mainly from gossans that were residually enriched in gold by the weathering of sulfides.

Most of the deposits are in the form of quartz veins, but one, at the Queen of Sheba mine, is a sulfide-bearing tactite. Some veins are in slate on or near contacts with serpentinized ultramafic rock or granitic bodies. Many occur in the serpentinite body, commonly on or near the contact with feldspar porphyry dikes; at least one occurrence of free gold along shear zones in serpentinite without other gangue in the vein has been reported. Auriferous veins in granitic rocks are less common, although both of the largest producers in the district, the Cummings (McKeen) and the Dewey mines, occurred in such an environment. Most of the veins strike northeast to north-northeast and dip moderately to steeply eastward. A few trends are in the northwest quadrant.

At the Queen of Sheba mine, the mineralized rock is a garnet-pyroxene tactite and pyroxene hornfels that occurs locally in a narrow north-south-trending belt of schist west of Callahan. The schist is part of the central metamorphic belt and lies between serpentinite on the east and a large granitic pluton on the west. Probably the tactite was formed by metasomatic replacement of marble or calcareous schist. The tactite is partly replaced and is transected by veinlets of pyrite and chalcopyrite which, by weathering, formed a thin soft surficial gossan that has been mined. The grade and amount produced, though probably small, are not known. Analysis of a chip sample collected along about 75 feet of the east wall of the opencut approximately parallel to the strike showed 0.16 ounces gold per ton and 0.05 percent copper.

YREKA–FORT JONES DISTRICT

Several districts within a radius of 20 miles of Yreka are included .here (pl. 2) as the Yreka–Fort Jones district because of similarity of their modes of occurrence. These districts are the Klamath River, Yreka, Humbug, Deadwood, and Oro Fino. Several mines, including the Eliza, Schroeder, Morrison and Carlock, and Indian Girl have recorded productions between $100,000 and $750,000 in gold; some, including the Golden Eagle, Mono, and Jillson, produced between $500,000 and $1 million in gold.

Most of the deposits are in rocks of the western Paleozoic and Triassic belt; some are in granitic plutons. The distribution of gold lodes forms a fairly well defined northeasterly trending zone whose southeast boundary is a belt of highly deformed schists that separates less deformed rocks of the western Paleozoic and Triassic belt from a serpentinite belt and rocks of the eastern Paleozoic belt on the southeast. The northwest boundary is poorly defined, but the deposits gradually become less abundant in that direction and are absent from the metamorphic rocks of amphibolite facies which lie between a belt of serpentinite bodies and granitic intrusives and a large area of greenschist facies rocks. To the southwest the belt terminates against a fault and amphibolite facies rocks. Overlapping sedimentary rocks of Cretaceous age limit the mineralized zone on the northeast.

The country rocks of the mineralized belt, aside from the granitic intrusions, are predominantly basaltic greenstones and lesser amounts of interbedded sedimentary rocks, including dark fine-grained argillite and slaty argillite, chert, and recrystallized limestone, or marble. The intensity of metamorphism increases northwestward; the mines and prospects are most abundant in the least metamorphosed part of the terrane.

Most lode deposits of the districts in the vicinity of Yreka are in metavolcanic rocks. None is positively known to have been in metasedimentary rocks, although sheared argillite occurs on some of the dumps. A few are on or near contacts between argillite and greenstone.

The lodes were simple quartz veins containing free gold and pyrite, possibly auriferous. The quartz is milky, commonly with a few small open vugs, and may contain fragments of country rock. Calcite is plentiful as veinlets and inclusions in the quartz and as veins, veinlets, and irregular replacements in the country rock. Some quartz veins are also accompanied by chlorite. Sulfide minerals, mainly pyrite and less commonly chalcopyrite, may occur with the quartz and in the country rock near the vein, but generally sulfides are not plentiful.

Alteration of the country rock is neither intense nor widespread. Mainly it is within a few inches of the veins and consists of mild sericitization of feldspar, partial replacement by calcite, veining by chlorite, and alteration of titaniferous magnetite to leucoxene.

HAPPY CAMP DISTRICT

The Happy Camp district (pl. 2) is in western Siskiyou County in the vicinity of Happy Camp, a town on the Klamath River approximately 40 airline miles west of Yreka. A few scattered prospects and mines which produced a small amount of gold are known. The district is noteworthy mainly because of a nearby copper deposit, the Gray Eagle, from which there has been substantial production, and, more pertinent to this report, a single gold deposit at the Siskon mine from which well over $3 million was produced between 1953 and 1961. The Siskon is, therefore, one of the largest, and probably the most recently productive, gold mines in the Klamath Mountains. In addition to the one at Gray Eagle mine, several other small copper sulfide deposits are known. The district is part of a larger region of copper mineralization in northwestern California and southwestern Oregon.

The deposits of the Happy Camp district are far west of the main gold-producing belt of the California part of the Klamath Mountains province. The geology of the region is known in only the most general terms, but this district is in the western part of the belt of western Paleozoic and Triassic rocks near the boundary with the western Jurassic belt. Rocks of both belts are intruded by serpentinized ultramafic rocks, gabbro, and plutons of dioritic to quartz dioritic composition. The deposits which have been prospected for gold are principally tabular sulfide replacement bodies accompanied by small intrusive bodies of gabbro and diorite. Weathering of the auriferous sulfide zone has resulted in the formation of ferruginous gossan or iron capping at the surface and enrichment in gold. None except the Siskon has been mined profitably for gold. Only the Siskon mine was visited during the present study; however, the other gossan deposits in this group are probably similar.

On the geologic map (pl. 1) the Siskon deposit can be seen to be located in the western Jurassic belt; however, because of the lack of detailed mapping, the bedrock near the mine is not well known. Rocks which are thought to be an outlier of the western Paleozoic and Triassic belt crop out nearby. A half mile east of the mine, however, there are slate, phyllite, and minor limestone units which are probably assignable to the Galice Formation of Jurassic age.

The mineralized zone, averaging approximately 50 feet but as much as 200 feet in width and 3,000 feet in length, strikes about N. 10° E. and dips 65°–70° W. Its surface trace is marked by a gossan, unmined parts of which remain at the north and south ends of the mine as reddish-brown bold knobs and pinnacles composed of porous limonite-cemented breccia. The intervening part of the gossan has been mined in open pits. The unweathered mineralized rock is a greenish-gray microgranular quartzite charged with abundant fine-grained pyrite which replaces and veins the quartzite. The wallrock is fine-grained slaty and siliceous metasedimentary and siliceous metavolcanic rocks. A small stock of gabbro intrudes the wallrock northeast of the mine; a few hundred feet west of the ore zone is the south end of a large granitic pluton.

LODE GOLD DISTRICTS IN OREGON
ASHLAND DISTRICT

Several small deposits and two which produced a significant amount of gold are clustered about the northwest border of the Ashland granitic pluton, 2–5 miles west and southwest of the town of Ashland (pl. 2). Most of the discoveries were on small veins from which no more than a few thousand dollars worth of gold were produced. The Ashland mine, however, had an estimated gold production of $1.3 million, and the Shorty Hope mine produced approximately $30,000.

Some deposits are within granitic rocks, and some are in metavolcanic and metasedimentary rocks of the western Paleozoic and Triassic belt near the contact of the granitic pluton. The deposits are in simple quartz-filled fractures that strike northeast and northwest and have moderate to steep dips. Metallic minerals of the veins, in addition to gold, include pyrite, chalcopyrite, and pyrrhotite. The gold is mostly free milling. The veins at most of the deposits seem to have been small and discontinuous. The Ashland mine, however, was developed in granitic rocks on a persistent, quartz-filled fracture that strikes N. 15°–20° E. and dips 40°–45° SE. The Shorty Hope mine, approximately 1 mile west of Ashland, was on a persistent northwest-striking vein in contact-metamorphosed rocks of the western Paleozoic and Triassic belt adjacent to the granitic pluton, offshoots of which complexly intrude the country rocks.

ROGUE RIVER–APPLEGATE DISTRICT

The Rogue River–Applegate district (pl. 2), as used here, forms a large triangle of land mainly in Jackson County and the eastern part of Josephine County and in the drainage of the Rogue and Applegate Rivers; this area includes several informally named but well-known mining districts (Winchell, 1914), including the Jacksonville, Gold Hill, and parts of the Upper and Lower Applegate districts and the Grants Pass district. There are approximately 150 known and reported small mines and prospects in the district. Production from most of the properties has amounted to no more than a few thousand dollars in gold; the Sylvanite and Opp mines, however, produced an estimated $700,000 and more than $100,000, respectively, and two pocket mines, the Gold Hill and Revenue, yielded $700,000 and $100,000, respectively.

The Rogue River–Applegate district is underlain by rocks of the western Paleozoic and Triassic belt, called the Applegate Group in southwestern Oregon (Wells and others, 1949, p. 3–4). The Applegate Group in this area is composed dominantly of metavolcanic rocks with lesser amounts of interbedded lenticular metasedimentary rocks, including argillite, chert, quartzite, and minor limestone. These rocks are intruded by several plutons of widely varying size and composition—granodiorite to gabbro. Most of the deposits are in rocks of the Applegate Group, but some are in the intrusive bodies. The widespread distribution of intrusive granitic rocks suggests that plutonic bodies lie at relatively shallow depths throughout the area.

The many mines and prospects in this district have shown that the gold occurs in small nonpersistent quartz veins and veinlets and in brecciated zones. The country rock adjacent to the veins is essentially barren. Many of the discoveries in this area were pockets which yielded relatively large amounts of free gold. The veins and veinlets have no clearly dominant orientation. Of the reported vein attitudes (Oregon Dept. Geology and Mineral Industries, 1943, 1952), however, west and northwest strikes and north to northeast and south to southeast dips are most common. The next most frequently reported attitudes are northeast strikes with northwest and southeast dips.

Free-milling gold occurs in the quartz and also in pyrite, which is the most common sulfide mineral. Other sulfides which may be present in distinctly minor quantities are chalcopyrite, pyrrhotite, and galena; sphalerite and gold telluride have also been reported. Calcite is a fairly common late-stage gangue mineral.

Analyses of samples of a mineralized fracture zone in siliceous schist in the valley of Evans Creek, Oreg., north of the Rogue River are contained in table 5.

ILLINOIS RIVER DISTRICT

Included in the Illinois River district (pl. 2) are the old Illinois River and Chetco districts in southwestern Josephine and eastern Curry Counties, Oreg. The area is in rugged, rather inaccessible terrain west and southwest of Selma, Oreg., and is mostly south of the Illinois River in the drainages of the Illinois River and the headwaters of the Chetco River. Production from the district has been negligible and has come mostly from small relatively high grade pockets, several of which yielded as much as several thousand dollars worth of gold. The Robert E. mine is estimated (Shenon, 1933a, p. 52) to have yielded more than $100,000 in gold, mostly from a restricted high-grade part of the vein.

In general, the Illinois River district is in a peninsulalike southwestward extension of the western Jurassic belt into serpentinites of the Josephine ultramafic pluton. The country rock in which the gold-bearing lodes occur is commonly metavolcanic rocks belonging to the Rogue Formation. Several deposits have been reported as being on or near the contact between metavolcanic rocks and serpentinite. A few have been described as occurring in serpentinite.

The deposits occur in quartz veins and in shear zones with which quartz veins are associated. The veins are small, irregular, and discontinuous. Gold occurred as free metal in quartz. An apparently typical feature of the Illinois River area is the common occurrence of sulfide minerals with the gold-bearing lodes. Pyrite, pyrrhotite, and chalcopyrite are usually reported to be present, commonly in abundance. Arsenopyrite is reported at some places. Pyrite and chalcopyrite commonly are gold bearing, and several of the original discoveries of gold were in weathered parts of sulfide-bearing veins. Some of the weathered veins may account for the high-grade pockets that were mined in shallow parts of the veins. Gold tellurides, unidentified as a mineral species, are also reported.

Shenon (1933a, p. 51) reported on the only deposit that yielded a significant past production, the Robert E. mine. Most of the gold produced came from a small shoot of partly oxidized sulfide ore in a quartz vein in greenstone. The ore shoot was near the contact of greenstone with serpentinite. Although the vein, which strikes slightly north of east and dips steeply south, was followed underground for as much as 300 feet, no other high-grade shoots were discovered.

TAKILMA DISTRCT

The Takilma district (pl. 2) is in southern Josephine County near the State boundary and the village of Takilma in the southern part of the valley of the East Fork of the Illinois River. The area has been chiefly a copper-producing district but was also noted for its placer gold deposits. It has had a small, unimportant lode gold production. The copper deposits and gold placers were described by Shenon (1933b).

The Takilma district is underlain by metavolcanic and fine-grained metasedimentary rocks of the western Paleozoic and Triassic belt and several small irregularly shaped serpentinite plutons. Included with the metavolcanic rocks are metagabbro bodies of undetermined size and distribution (Shenon, 1933b, p. 158; Wells and others, 1940, 1949). Granitic plutons are generally absent, except for a few small dikes and plugs.

Irregularly shaped copper deposits, mostly in greenstone, have been found near contacts of serpentinite with metavolcanic rocks and metagabbro (Shenon, 1933b, p. 162). These deposits are distributed in a narrow north-south belt approximately 6 miles long east of Takilma. The copper ores are sulfides, including chalcopyrite, cubanite, sphalerite, pyrite, and pyrrhotite, which were deposited in fractures and to a lesser extent as disseminations in wallrocks adjoining fractures. Shenon (1933b, p. 162) reported a production from the copper deposits of approximately $1.7 million. Gold content of the deposits is low, ranging from about 0.04 to 0.1 ounce per ton, and would be recoverable only as a byproduct under conditions that permitted the profitable mining of the copper ores.

The gold-bearing lodes are mostly east of the Takilma copper district in the drainages of Sucker and Althouse Creeks. All the deposits are small; several yielded a few thousand dollars worth of gold from high-grade pockets and were abandoned when further exploration failed to discover more ore. The deposits are in steeply dipping quartz veins a few inches to as much as 5 feet wide that strike northwest or northeast. The gold occurs free and is commonly accompanied by some sulfide minerals, mainly pyrite and minor amounts of chalcopyrite and galena. The veins occupy fractures in both metavolcanic and metasedimentary rocks; some are at contacts with serpentinite. The rich placer deposits in Althouse and Sucker Creeks undoubtedly derived their gold from this region, probably from small discontinuous veins and pocket deposits similar to those described.

GREENBACK DISTRICT

The Greenback district (pl. 2) is approximately 15 miles north of Grants Pass in northeastern Josephine, northwestern Jackson, and southern Douglas Counties and is named for the Greenback mine, the largest producer in the area. For the most part it lies in the upper drainages of Grave, Coyote, and Wolf Creeks. Several small mines and prospects produced from a few thousand dollars to a few tens of thousand dollars in gold; the Daisy or Hammersley mine probably produced about $250,000 worth of gold, and the Greenback mine, approximately $3.5 million.

The Greenback district has several geologic features which distinguish it from other districts in southwestern Oregon. It is on a northeast-trending zone of serpentinite which marks the boundary between rocks of the western Paleozoic and Triassic belt and the western Jurassic belt. An inlier of the Galice Formation is faulted against rocks of the Applegate Group and has discontinuous small serpentinite bodies along the contact; these serpentinites are not shown on plate 1. Reconnaissance studies have shown also that many bodies of fine-grained gabbro intrude rocks of both the western Jurassic and the western Paleozoic and Triassic belts. Gabbro was not distinguished from the metavolcanic rocks by Diller and Kay (1924), and because its distribution is not clearly known from the reconnaissance work, it is not shown on plate 1. According to the present interpretation of the geology, the serpentinite and gabbro were intruded into a zone of thrust faults between the eastern Paleozoic and Triassic belt and the western Jurassic belt. Another feature that distinguishes this area is the absence of granitic rocks, which are so common in some of the other mineralized areas of southeastern Oregon; however, large granitic plutons intrude mafic and ultramafic rocks along the contact zone to the northeast and southeast.

Most of the mines, including the Greenback and the Daisy or Hammersley, are in greenstone in a narrow zone intruded by serpentinite and gabbro adjacent to the contact with sedimentary rocks of the Galice Formation. A few small mines and prospects are in the Galice close to the contact. Some small isolated deposits are known in greenstone outside the main part of the district. In many mines the gold-bearing veins are reported to have nearly east strikes and moderate to steep northward dips (Diller and Kay, 1908; Winchell, 1914; Oregon Dept. Geology and Mineral Industries, 1952). The veins range from narrow seams to veins approximately 4 feet in width, but generally they are less than 1 foot in width. The veins are irregular and discontinuous and further complicated by faulting.

Most of the gold occurred free in quartz veins, though it was also contained in pyrite, which is commonly present in the veins and in the adjacent country rock. Of other sulfide minerals in the veins, chalcopyrite is the most commonly reported, though in minor amounts; pyrrhotite and galena are rarely reported; arsenopyrite is the most uncommon. Calcite is a fairly common nonmetallic gangue mineral. At a few places gossan formed by weathering of pyritiferous veins was mined in surficial workings for its gold content.

Analyses of samples from an exploration cut in greenstone at the Greenback mine are contained in table 6.

GALICE DISTRICT

The Galice district, as used here, is a narrow elongate area in northern Josephine County (pl. 2) which embraces the Galice and Mount Reuben districts. It is approximately 20 miles northwest of Grants Pass and is bisected by the Rogue River. The area was one of the richest producers in southwestern Oregon and, for its size, has a high concentration of mines and prospects. Of approximately 80 known mines and prospects (Oregon Dept. Geology and Mineral Industries, 1952; Youngberg, 1947), most had known productions of a few thousand dollars in gold. The largest producers and their approximate yields were the J.C.L. ($100,000), Robertson or Bunker Hill ($138,-000), Gold Bug ($750,000), and Benton ($500,000–$800,000); the Almeda mine was chiefly a copper producer but yielded about $31,000 in gold.

The mineralized belt is approximately 5 miles wide and 20 miles long and is on the northwest border of the Klamath Mountains geologic province. It trends north-northeast, parallel to the regional strike of the rocks, and is bounded on the east by slate of the Galice Formation and on the west by sedimentary rocks of the Dothan Formation. Country rocks of the mineralized belt are metavolcanics of the Rogue Formation (Wells and Walker, 1953) and rocks of the almandine-amphibolite metamorphic facies that Wells and Walker (1953) concluded were metamorphic equivalents of the Rogue Formation. The rocks are cut by three major northeast-trending faults along which there are narrow continuous bodies of serpentinite. Two of these faults are apparently associated with a fracture zone which is complexly intruded by bodies of gabbro and granitic rock.

The northern part of the Galice district, north of the Rogue River, is usually referred to as the Mount Reuben district and was studied in considerable detail by Youngberg (1947). Adjacent to the contact with the Dothan Formation, the Rogue Formation is complexly intruded by gabbro, quartz diorite, and some serpentinite. East of the intrusive zone are greenstones of the Rogue Formation.

Youngberg (1947, p. 8–9) called attention to differences in the nature of mineralization, depending on the kind of country rock in which the deposits occur.

Veins in the gabbro complex have been essentially nonproductive. Although they are fairly persistent, their gold content is generally low. Chalcopyrite and pyrite are the principal sulfide minerals, and pyrrhotite is common.

Deposits in the greenstone occur in relatively short narrow ore shoots along well-defined shear zones. Usually no more than one small shoot has been discovered along a given vein, which limits mining to a small-scale operation. Gold occurs free in quartz and associated with pyrite and chalcopyrite.

The most productive veins were in a small quartz diorite body at the Benton mine, where persistent veins containing ore shoots as much as several hundred feet long were found in a system of intersecting fractures. Besides quartz-filled fissures there is also considerable sheared and altered quartz diorite. Pyrite occurs in the quartz and altered granitic rock in and adjacent to the veins. The gold is mostly contained in pyrite, and there is little free metal. Some molybdenite is present; chalcopyrite and pyrrhotite are very rare.

The southern part of the Galice district is mainly east of the gabbro intrusive complex. Most of the mines and prospects are in a belt of amphibolite-grade metamorphic rocks that lies between a narrow wedge of metavolcanic rocks of the Rogue Formation and the mafic intrusive complex. A few are in Rogue Formation greenstones, and a few are in gabbroic rocks of the complex. Deposits in the greenstone and metamorphic rocks are in veins along fractures that most commonly have a northeast strike. The vein fillings are quartz, invariably accompanied by pyrite and minor amounts of chalcopyrite, and rarely, galena and sphalerite. Small amounts of free gold occur in the quartz, but gold is mostly associated with pyrite. With the exception of the Robertson or Bunker Hill mine, none of the deposits has produced more than a few thousand dollars worth of gold, mainly because of the small and discontinuous nature of the veins.

The Robertson or Bunker Hill mine is west of the zone of complex fracturing and mafic intrusives, at the north end of a tongue of metavolcanic rocks enclosed on three sides by quartz diorite (Shenon, 1933a, p. 40–48). The deposit consists of a group of four northeast-striking, southwest-dipping veins. Like others in this area, the veins are no more than 100–200 feet long and are highly variable in width. Gold occurs free in quartz, in pyrite, and in a gold-silver telluride, petzite.

The Almeda mine, also in the Galice district, is a large sulfide replacement deposit containing copper, gold, and silver (Diller, 1914, p. 72–81; Shenon, 1933c, p. 24–35). Recently Libbey (1967) reviewed its history and gathered previously unavailable data on past exploration.

The deposit is on the Rogue River at the contact between the Galice and Rogue Formations. This contact, which is a fault, has been the site of a sill-like intrusion of porphyritic dacite; similar tabular bodies of porphyritic dacite intrude rocks on both sides of the contact, according to Shenon (1933c, p. 26). The porphyritic dacite and the greenstone adjacent to the contact have been partly to completely replaced by fine-grained quartz. The silicified rocks in turn have been replaced to varying degrees by sulfides and barite. Shenon (1933c, p. 30) reported that the sulfides include pyrite, which is by far the most abundant, chalcopyrite, galena, and sphalerite. Most of the quartz preceded deposition of barite, which was followed by introduction of the sulfides.

Assays of samples collected underground and by diamond drilling (Libbey, 1967, p. 18) indicate that the ore ranges in grade from 0.057 to 0.155 ounce gold per ton, 2.49 to 4.34 ounce silver per ton, and 0.71 to 1.25 percent copper. Gold is presumably contained mostly in pyrite, for some of the highest gold values were obtained from pyritic siliceous mineralized rock low in chalcopyrite.

Shenon (1933c, p. 33) concluded that there is a large deposit of mineralized rock at the Almeda mine which, after careful sampling and exploration to determine its extent, might be minable under favorable conditions. Exploration by more than 7,000 feet of underground workings and approximately 24 diamond-drill holes (Libbey, 1967) indicates that the ore is fairly persistent with depth, although insufficient work has been done to clearly determine the extent of the ore bodies.

The mineralized zone at the Almeda mine is in what Diller (1914, p. 74) mentioned as being generally known in the region as the Big Yank lode. He called attention to its occurrence at the contact between slates of the Galice Formation and volcanic rocks of the Rogue Formation and said that the contact could be followed for over 20 miles from Cow Creek on the north to Briggs Creek Valley on the south. However, Dole and Baldwin (1947), as a result of reconnaissance north of the Almeda, found no evidence of mineralization that was clearly related to this contact. To the southeast some evidence of alteration and mineralization can be seen in roadcuts. In sec. 26, T. 4 S., R. 8 W., where the contact crosses the road up Rocky Gulch to the Oriole mine, a zone of altered, bleached, and silicified greenstone 50–75 feet wide is exposed. Some very fine grained pyrite is visible in some specimens. In Sailor Jack Creek in the NW¼ sec. 10, T. 35 S., R. 8 W., on Galice Creek road, a zone of altered, bleached, and iron-stained rock 10 feet wide is exposed at the contact between slate and greenstone. Outcrops of the mineralized zone exposed in the bed of Rogue River and in roadcuts were sampled during this investigation (fig. 6). Analytical data are contained in table 7.

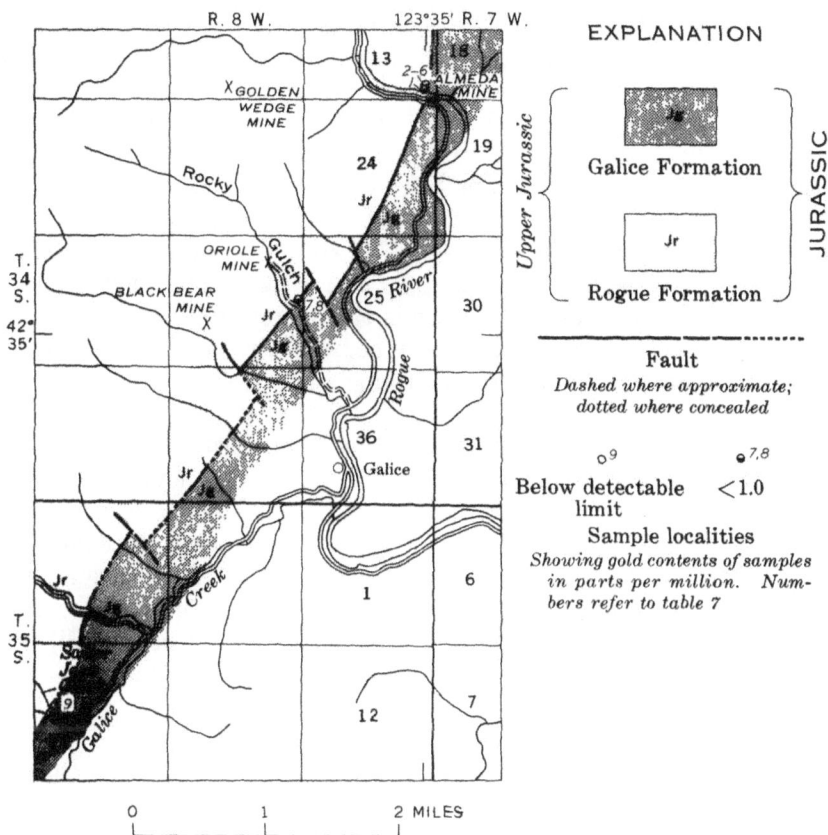

FIGURE 6.—Sample localities, Galice district, Oregon.

SILVER PEAK DISTRICT

The Silver Peak (or Silver Butte) district is in the northeastern part of the Klamath Mountains province in southern Douglas County. It is a narrow mineralized zone 2–8 miles southwest of the town of Canyonville (pl. 2). The principal mines, those of the Silver Peak Copper Co., the Umpqua Consolidated Mining Co., and the Golden Gate mine, were described by Shenon (1933c, p. 15–24), who estimated that the total production was worth approximately $73,000. Most of the value lay in copper, with minor amounts of gold and silver.

The Silver Peak district is of interest because its deposits have several features in common with that at the Almeda mine in the Galice area. The deposits of the Silver Peak district are principally copper-bearing sulfide deposits with minor zinc and subordinate gold and silver and, like the Almeda deposit, are characterized by barite gangue in addition to quartz. They occur as replacement deposits along narrow shear zones in metavolcanic rocks. In contrast to the Almeda deposit, which is in greenstone on the southeast side of the belt of Rogue Formation adjacent to the contact with sedimentary rocks of the Galice Formation, the Silver Peak deposits are in schistose metavolcanic rocks on the northwest side of the belt near the contact with sedimentary rocks of the Dothan Formation. The mines and prospects are alined parallel to and approximately one-half mile southeast of the contact, which strikes approximately N. 45° E. The Dothan-Rogue contact is a moderate to steep southeast dipping fault. Narrow serpentinite bodies occur at several places along the fault and in the metavolcanic rocks near the mineralized zone.

The sulfide minerals include pyrite, sphalerite, chalcopyrite, bornite, and relatively small amounts of galena, tennantite, chalcocite, and covellite. Gangue minerals are mainly quartz, barite, and sericite (Shenon, 1933c, p. 18). Gold, which presumably is in the sulfide minerals, assayed 0.01–0.09 ounce per ton in samples collected by Shenon (1933c, p. 20) and from 0.04 to 0.24 ounce per ton with from 2.2 to 7.3 ounces per ton of silver in ore mined from the Silver Peak and Umpqua Consolidated mines (Shenon, 1933c, p. 21).

MYRTLE CREEK DISTRICT

The Continental and Chieftain mines are 12 miles northeast of Canyonville on South Myrtle Creek (pl. 2). They occur alone in an area that has little or no mineralization, although placer gold has been mined to the north in the headwaters of North Myrtle Creek. They were briefly studied and described by Wells (1933). Little is known about their production. The mines were in operation from approximately 1918 to 1935; apparently most of the production was from the Chieftain mine.

The geology of the region surrounding the Continental and Chieftain mines is poorly known. The country rocks are metavolcanics which have been intruded by gabbroic bodies of unknown size (Diller, 1898; Wells, 1933; Wells and Peck, 1961); the metavolcanic rocks are part of the Galice-Rogue assemblage. The more or less altered gabbroic rocks are commonly composed of secondary hornblende, chlorite, and altered plagioclase. Where seen by the author, the gabbro is commonly highly fractured and is cut by intersecting white veinlets of laumontite.

According to Wells (1933), the Chieftain and Continental mines were on a nearly east-striking vein that dips steeply north in altered gabbroic rock. The vein has been traced along strike at the surface for approximately 1½ miles and has been followed on strike underground at the Chieftain for 640 feet and at the Continental for 500 feet. The vein consists of lenses and discontinuous stringers of quartz in a shear zone as much as 4 feet wide. Metallic minerals occurring in the shear zone and associated with quartz are predominantly pyrite with lesser amounts of chalcopyrite and sphalerite; in addition, the gold and silver tellurides, sylvanite and petzite, occur in the chalcopyrite and sphalerite, and in quartz. No free gold is reported. Wells (1933) concluded that the mineralized vein probably continues in depth and a considerable reserve tonnage of ore remains.

MULE CREEK DISTRICT

The Mule Creek district is in an isolated rugged part of northeastern Curry County near the settlement of Marial on the Rogue River approximately 35 miles northwest of Grants Pass (pl. 2). The Mule Creek drainage is north of the river, and Mule Creek enters the river at Marial. The area of mineralization is small and relatively unimportant economically, but it is of interest because of its isolated and seemingly anomalous position west of the main Klamath Mountains mineralized region. A belt of apparently unmineralized rocks of the Dothan Formation approximately 15 mile wide separates the Mule Creek district from the Galice district.

The geology of the Mule Creek district is poorly known. Gold mineralization is confined to a narrow belt of metavolcanic rocks and mafic intrusive rocks that lies west of the sedimentary rocks of the Dothan Formation near the overlap of Cretaceous and Tertiary marine deposits of the Oregon Coast Range. Wells (1955) and Wells and Peck (1961) assigned the metavolcanic rocks to the Dothan Formation. These metavolcanic rocks are intruded by diabasic dikes; a body of quartz gabbro which is exposed in the Rogue River canyon approximately 2 miles southwest of Marial may also intrude the volcanics, although its west boundary at the river level appears to be faulted. The east limit of the volcanic rocks is certainly in part a fault. If the Dothan Formation is younger than the episodes of intrusion and gold mineralization elsewhere in the Klamath Mountains, the greenstones of the Mule Creek district must be older and possibly equivalent to one of the older volcanic units, perhaps the Rogue or Galice Formations.

The Mule Creek deposits occur in veins in the metavolcanic rocks. The veins, which are generally small and discontinuous, consist of quartz and minor amounts of pyrite and chalcopyrite. Occasional high-grade pockets were mined from these veins. Most of the gold mined seems to have occurred free in the quartz, although Butler and Mitchell (1916) reported that some lower grade lodes were mined in which much of the gold was in chalcopyrite and pyrite along shear zones containing quartz stringers.

REFERENCES CITED

Albers, J. P., 1961, Gold deposits in the French Gulch–Deadwood district, Shasta and Trinity Counties, California, in Geological Survey research, 1961: U.S. Geol. Survey Prof. Paper 424–C, p. C1–C4.

———— 1964, Geology of the French Gulch quadrangle, Shasta and Trinity Counties, California: U.S. Geol. Survey Bull. 1141–J, p. J1–J70.

———— 1965, Economic geology of the French Gulch quadrangle, Shasta and Trinity Counties, California: California Div. Mines and Geology Spec. Rept. 85, 43 p.

———— 1966, Economic deposits of the Klamath Mountains, in Bailey, E. H., ed., Geology of northern California: California Div. Mines and Geology Bull. 190, p. 51–62.

Albers, J. P., Kinkel, A. R., Jr., Drake, A. A., and Irwin, W. P., 1964, Geology of the French Gulch quadrangle, California: U.S. Geol. Survey Geol. Quad. Map GQ–336, scale 1 : 62,500.

Alberts, J. P., and Robertson, J. F., 1961, Geology and ore deposits of east Shasta copper-zinc district, Shasta County, California: U.S. Geol. Survey Prof. Paper 338, 107 p.

Averill, C. V., 1933, Gold deposits of the Redding and Weaverville quadrangles: California Jour. Mines and Geology, v. 29, nos. 1–2, p. 2–73.

———— 1935, Mines and mineral resources of Siskiyou County: California Jour. Mines and Geology, v. 31, no. 3, p. 255–338.

———— 1939, Mineral resources of Shasta County: California Jour. Mines and Geology, v. 35, no. 2, p. 108–191.

Blake, M. C., Jr., Irwin, W. P., and Coleman, R. G., 1967, Upside-down metamorphic zonation, blueschist facies, along a regional thrust in California and Oregon, in Geological Survey research, 1967: U.S. Geol. Survey Prof. Paper 575–C, p. C1–C9.

Brown, G. C., 1916, The counties of Shasta, Siskiyou, Trinity: California Mining Bur., 14th Rept. State Mineralogist, p. 745–925.

Butler, G. M., and Mitchell, G. J., 1916, Preliminary survey of the geology and mineral resources of Curry County, Oregon: Oregon Bur. Mines and Geology, Mineral Resources Oregon, v. 2, no. 2, 134 p.

Cater, F. W., Jr., and Wells, F. G., 1953, Geology and mineral resources of the Gasquet quadrangle, California-Oregon: U.S. Geol. Survey Bull. 995–C, p. 79–133.

Churkin, Michael, Jr., 1965, First occurrence of graptolites in the Klamath Mountains, California, *in* Geological Survey research, 1965: U.S. Geol. Survey Prof. Paper 525–C, p. C72–C73.

Churkin, Michael, Jr., and Langenheim, R. L., Jr., 1960, Silurian strata of the Klamath Mountains, California: Am. Jour. Sci., v. 258, no. 4, p. 258–273.

Cloke, P. L., and Kelly, W. C., 1964, Solubility of gold under inorganic conditions: Econ. Geology, v. 59, no. 2, p. 259–270.

Cox, D. P., 1956, Geology of the Helena quadrangle, Trinity County, California: Stanford, Calif., Stanford Univ. Ph.D. thesis.

Curtis, G. H., Evernden, J. F., and Lipson, J. I., 1958, Age determination of some granitic rocks in California by the potassium-argon method: California Div. Mines Spec. Rept. 54, 16 p.

Davis, G. A., 1963, Structure and mode of emplacement of Caribou Mountain pluton, Klamath Mountains, California: Geol. Soc. America Bull., v. 74, no. 3, p. 331–348.

———— 1965, Regional Mesozoic thrusting in the south-central Klamath Mountains of California [abs.]: Geol. Soc. America Spec. Paper 82, p. 248.

———— 1966, Metamorphic and granitic history of the Klamath Mountains, *in* Bailey, E. H., ed., Geology of northern California: California Div. Mines and Geology Bull. 190, p. 39–50.

———— 1968, Westward thrust faulting in the south-central Klamath Mountains, California: Geol. Soc. America Bull., v. 79, no. 7, p. 911–834.

Davis, G. A., Holdaway, M. J., Lipman, P. W., and Romey, W. D., 1965, Structure, metamorphism, and plutonism in the south-central Klamath Mountains, California: Geol. Soc. America Bull., v. 76, no. 8, p. 933–966.

Davis, G. A., and Lipman, P. W., 1962, Revised structural sequence of pre-Cretaceous metamorphic rocks in the southern Klamath Mountains, California: Geol. Soc. America Bull., v. 73, no. 12, p. 1547–1552.

Diller, J. S., 1886, Notes on the geology of northern California: U.S. Geol. Survey Bull. 33, 23 p.

———— 1898, Description of the Roseburg quadrangle [Oregon]: U.S. Survey Geol. Atlas, Folio 49.

———— 1902, Topographic development of the Klamath Mountains: U.S. Geol. Survey Bull. 196, 69 p.

———— 1903, Klamath Mountain section, California: Am. Jour. Sci., 4th ser., v. 15, p. 342–362.

———— 1906, Description of the Redding quadrangle, California: U.S. Geol. Survey Geol. Atlas, Folio 138, 14 p.

———— 1914, Mineral resources of southwestern Oregon: U.S. Geol. Survey Bull. 546, 147 p.

———— 1922, Chromite in the Klamath Mountains, California and Oregon: U.S. Geol. Survey Bull. 725, p. 1–35.

Diller, J. S., and Kay, G. F., 1908, Mines of the Riddle quadrangle, Oregon: U.S. Geol. Survey Bull. 340–A, p. 134–152.

———— 1924, Description of the Riddle quadrangle [Oregon]: U.S. Geol. Survey Geol. Atlas, Folio 218, 8 p.

Dole, H. M., and Baldwin, E. M., 1947, A reconnaissance between the Almeda and Silver Peak mines of southwestern Oregon: Ore Bin, v. 9, no. 12, p. 95–100.

Ferguson, H. G., 1914, Gold lodes of the Weaverville quadrangle, California: U.S. Geol. Survey Bull. 540, p. 22–79.

———— 1915, Pocket deposits of the Klamath Mountains, California: Econ. Geology, v. 10, p. 241–261.

Helgeson, H. C., and Garrels, R. M., 1968, Hydrothermal transport and deposition of gold: Econ. Geology, v. 63, no. 6, p. 622–635.

Hershey, O. H., 1899, The upper Coffee Creek mining district [California]: Mining and Sci. Press, v. 79, p. 689.

———— 1900, Gold-bearing lodes of the Sierra Costa Mountains in California: Am. Geologist, v. 25, p. 76–96.

———— 1901, Metamorphic formations of northwestern California: Am. Geologist, v. 27, p. 225–245.

———— 1910, Origin of gold "pockets" in northern California: Mining and Sci. Press, v. 101, no. 23, p. 741–742.

Hinds, N. E. A., 1932, Paleozoic eruptive rocks of the southern Klamath Mountains, California: California Univ., Dept. Geol. Sci. Bull., v. 20, no. 11, p. 375–410.

———— 1933, Geologic formations of the Redding-Weaverville districts, northern California: California Jour. Mines and Geology, v. 29, p. 76–122.

———— 1935, Mesozoic and Cenozoic eruptive rocks of the southern Klamath Mountains, California: California Univ., Dept. Geol. Sci. Bull., v. 23, no. 11, p. 313–380.

Hollister, V. F., and Evans, J. R., 1965, Geology of the Redding quadrangle, Shasta County, California: California Div. Mines and Geology Map Sheet Ser. 4, scale 1 : 24,000.

Hotz, P. E., 1967, Geologic map of the Condrey Mountain quadrangle and parts of the Seiad Valley and Hornbrook quadrangles [California]: U.S. Geol. Survey Geol. Quad. Map GQ–618, scale 1 : 26,500.

Irwin, W. P., 1960a, Relations between Abrams mica schist and Salmon hornblende schist in Weaverville quadrangle, California, in Geological Survey research, 1960: U.S. Geol. Survey Prof. Paper 400–B, p. B315–B316.

———— 1960b, Geological reconnaissance of the northern Coast Ranges and Klamath Mountains, California, with a summary of the mineral resources: California Div. Mines Bull. 179, 80 p.

———— 1963, Preliminary geologic map of the Weaverville quadrangle, California: U.S. Geol. Survey Mineral Inv. Field Studies Map MF–275, scale 1 : 62,500.

———— 1964, Late Mesozoic orogenies in the ultramafic belts of northwestern California and southwestern Oregon, in Geological Survey research, 1964: U.S. Geol. Survey Prof. Paper 501–C, p. C1–C9.

———— 1966, Geology of the Klamath Mountains province, in Bailey, E. H., ed., Geology of northern California: California Div. Mines and Geology Bull. 190, p. 19–38.

Irwin, W. P., and Lipman, P. W., 1962, A regional ultramafic sheet in eastern Klamath Mountains, California, in Geological Survey research, 1962: U.S. Geol. Survey Prof. Paper 450–C, p. C18–C21.

Jones, D. L., 1960, Lower Cretaceous (Albian) fossils from southwestern Oregon and their paleogeographic significance: Jour. Paleontology, v. 34, no. 1, p. 152–160.

Kinkel, A. R., Jr., and Albers, J. P., 1951, Geology of the massive sulfide deposits at Iron Mountain, Shasta County, California: California Div. Mines Spec. Rept. 14, 19 p.

Kinkel, A. R., Jr., Hall, W. E., and Albers, J. P., 1956, Geology and base-metal deposits of West Shasta copper-zinc district, Shasta County, California: U.S. Geol. Survey Prof. Paper 285, 156 p.

Knopf, Adolph, 1929, The Mother Lode system of California: U.S. Geol. Survey Prof. Paper 157, 88 p.

———— 1936, Igneous geology of the Spanish Peaks region, Colorado: Geol. Soc. America Bull., v. 47, no. 11, p. 1727–1784.

Krauskopf, K. B., 1951, The solubility of gold: Econ. Geology, v. 46, no. 8, p. 858–870.

LaFehr, T. R., 1966, Gravity in the eastern Klamath Mountains, California: Geol. Soc. America Bull., v. 77, no. 11, p. 1177–1190.

Lanphere, M. A., and Irwin, W. P., 1965, Carboniferous isotopic age of the metamorphism of the Salmon Hornblende Schist and Abrams Mica Schist, southern Klamath Mountains, California: U.S. Geol. Survey Prof. Paper 525–D, p. D27–D33.

Lanphere, M. A., Irwin, W. P., and Hotz, P. E., 1968, Isotopic age of the Nevada orogeny and older plutonic and metamorphic events in the Klamath Mountains, California: Geol. Soc. America Bull., v. 79, no. 8, p. 1027–1052.

———— 1969, Geochronology of crystalline rocks in the Klamath Mountains, California and Oregon [abs.]: Geol. Soc. America, Cordilleran Sec.–Paleont. Soc., Pacific Coast Sec., 65th Ann. Mtg., Eugene, Oreg., 1969, Programs, pt. 3, p. 34.

Leach, F. A., 1899, Fineness of California gold, in California mines and minerals: San Francisco, California Miners' Assoc., p. 175–187.

Libbey, F. W., 1963, Lest we forget: Ore Bin, v. 25, no. 6, p. 93–109.

———— 1967, The Almeda mine, Josephine County, Oregon: Oregon State Dept. Geol. and Mineral Industries Short Paper 24, 53 p.

Lipman, P. W., 1963, Gibson Peak pluton—A discordant composite intrusion in the southeastern Trinity Alps, northern California: Geol. Soc. America Bull., v. 74, no. 10, p. 1259–1280.

Logan, C. A., 1926, Gold (quartz mines), in Sacramento field division, Shasta County: California Mining Bur., 22d Rept. State Mineralogist, p. 167–190.

MacDonald, D. F., 1913, Notes on the gold lodes of the Carrville district, Trinity County, California: U.S. Geol. Survey Bull. 530, p. 9–41.

MacGinitie, H. D., 1937, The flora of the Weaverville beds of Trinity County, California; with descriptions of the plant-bearing beds, in Eocene flora of western America: Carnegie Inst. Washington Contr. Paleontology, Pub. 465, p. 83–151.

Maxson, J. H., 1933, Economic geology of portions of Del Norte and Siskiyou Counties, northwesternmost California: California Jour. Mines and Geology, v. 29, nos. 1–2, p. 123–160.

Medaris, L. G., 1966, Geology of the Seiad Valley area, Siskiyou County, California, and petrology of the Seiad ultramafic complex: Los Angeles, Calif., California Univ. Ph.D. thesis.

Merriam, C. W., 1961, Silurian and Devonian rocks of the Klamath Mountains, California, in Geological Survey research, 1961: U.S. Geol. Survey Prof. Paper 424–C, p. C188–C189.

Murphy, M. A., Peterson, G. L., and Rodda, P. U., 1964, Revision of Cretaceous lithostratigraphic nomenclature, northwest Sacramento Valley, California: Am. Assoc. Petroleum Geologists Bull., v. 48, no. 4, p. 496–502.

Oregon Department of Geology and Mineral Industries, 1943, Oregon metal mines handbook—Jackson County: Oregon Dept. Geology and Mineral Industries Bull., no. 14–C, v. 2, sec. 2, 208 p.

———— 1952, Oregon metal mines handbook—Josephine County [2d ed.]: Oregon Dept. Geology and Mineral Industries Bull., no. 14–C, v. 2, sec. 1, 238 p.

Peck, D. L., Imlay, R. W., and Popenoe, W. P., 1956, Upper Cretaceous rocks of parts of southwestern Oregon and northern California: Am. Assoc. Petroleum Geologists Bull., v. 40, no. 8, p. 1968–1984.

Poole, F. G., Baars, D. L., Drewes, H., Hayes, P. T., Ketner, K. B., McKee, E. D., Teichert, C., and Williams, J. S., 1967, Devonian of the southwestern United States, in Oswald, D. H., ed., International Symposium on the Devonian System, Calgary, 1967, v. 1: Calgary, Alberta, Alberta Society Petroleum Geologists, p. 879–912.

Raymond, R. W., 1874, Statistics of mines and mining in the states and territories west of the Rocky Mountains, 6th Annual Report: U.S. Treasury Dept., 585 p.

Romey, W. D., 1962, Geology of a part of the Etna quadrangle, Siskiyou County, California: Berkeley, Calif., California Univ. Ph.D. thesis, 93 p.

Rynearson, G. A., and Smith, C. T., 1940, Chromite deposits in the Seiad quadrangle, Siskiyou County, California: U.S. Geol. Survey Bull. 922–J, p. 281–306.

Sanborn, A. F., 1960, Geology and paleontology of the southwest quarter of the Big Bend quadrangle, Shasta County, California: California Div. Mines Spec. Rept. 63, 26 p.

Seyfert, C. K., Jr., 1964, Geology of the Sawyers Bar area, Klamath Mountains, northern California: Stanford, Calif., Stanford Univ. Ph.D. thesis.

Shenon, P. J., 1933a, Geology of the Robertson, Humdinger, and Robert E. gold mines, southwestern Oregon: U.S. Geol. Survey Bull. 830, p. 33–55.

————— 1933b, Geology and ore deposits of the Takilma-Waldo district, Oregon, including the Blue Creek district: U.S. Geol. Survey Bull. 846–B, p. 141–194.

————— 1933c, Copper deposits in the Squaw Creek and Silver Peak districts and at the Almeda mine, southwestern Oregon: U.S. Geol. Survey Circ. 2, 34 p.

Strand, R. G. (compiler), 1962, Geologic map of California, Redding sheet—Olaf P. Jenkins edition: California Div. Mines and Geology, scale 1: 250,000.

————— 1964, Geologic map of California, Weed sheet—Olaf P. Jenkins edition: California Div. Mines and Geology, scale 1: 250,000.

Thompson, H. M., 1957, King Solomon mine [Siskiyou County, California]: The Siskiyou Pioneer, v. 2, no. 10, p. 14–18.

Tucker, W. B., 1922, Silver lodes of the South Fork mining district, Shasta County: California Mining Bur., 18th Rept. State Mineralogist, p. 313–321.

U.S. Bureau of Mines, 1927–33, Mineral resources of the United States, 1924–31: U.S. Dept. Commerce, Bur. Mines.

————— 1933, Minerals Yearbook, 1932–33: U.S. Dept. Commerce, Bur. Mines.

————— 1934–67 [issued annually], Minerals Yearbooks, 1934–66: U.S. Dept. Interior, Bur. Mines.

U.S. Geological Survey, 1904–26, Minerals resources of the United States, 1902–23: U.S. Dept. Interior, Geol. Survey.

Wells, F. G., 1933, Notes on the Chieftain and Continental miles, Douglas County, Oregon: U.S. Geol. Survey Bull. 830, p. 57–62.

————— 1955, Preliminary geologic map of southwestern Oregon west of meridian 122° west and south of parallel 43° north: U.S. Geol. Survey Mineral Inv. Field Studies Map MF–38, scale 1: 250,000.

————— 1956, Geology of the Medford quadrangle, Oregon-California: U.S. Geol. Survey Geol. Quad. Map GQ–89, scale 1: 96,000.

Wells, F. G., Hotz, P. E., and Cater, F. W., 1949, Preliminary description of the geology of the Kerby quadrangle, Oregon: Oregon Dept. Geology and Mineral Industries Bull. 40, 23 p.

Wells, F. G., and Peck, D. L., 1961, Geologic map of Oregon west of the 121st medidian: U.S. Geol. Survey Misc. Map I–325, scale 1:500,000.

Wells, F. G., and Walker, G. W., 1953, Geologic map of the Galice quadrangle, Oregon: U.S. Geol. Survey Geol. Quad. Map GQ–25, scale 1:62,500.

Wells, F. G., Walker, G. W., and Merriam, C. W., 1959, Upper Ordovician (?) and Upper Silurian formations of the northern Klamath Mountains, California: Geol. Soc. America Bull., v. 70, no. 5, p. 645–649.

Wells, F. G., and others, 1940, Preliminary geologic map of the Grants Pass quadrangle, Oregon: Oregon Dept. Geology and Mineral Industries, scale 1:96,000.

Winchell, A. N., 1914, Petrology and mineral resources of Jackson and Josephine Counties, Oregon: Oregon Bur. Mines Mineral Resources, v. 1, no. 5, 170 p.

Youngberg, E. A., 1947, Mines and prospects of the Mount Reuben mining district, Josephine County, Oregon: Oregon Dept. Geology and Min. Industries Bull. 34, 35 p.

TABLES

TABLE 1. —*Analyses of samples,*

[Analyses, except gold, are semiquantitative spectrographic by E. E. Martinez and are symbols: N, not detected; L, detected but below limit of determination. Results are analyses by atomic absorption are by W. L. Campbell, R. L. Miller, M. S. Rickard, and

Sample No.	Laboratory No.	Field No.	Ag	As	Au	B	Ba	Be	Bi	Cd	Co	Cr	Cu	La
1	AEA—638	SM—1—67	N	N	N	50	300	1	N	N	15	100	50	N
2	639	2	L	N	N	20	150	L	N	N	20	30	50	N
3	640	3	L	N	N	100	1,500	5	N	N	70	70	100	N
Limits of determination			0.5	200	0.02	10	20	1	10	20	5	5	5	20

1. Roadcut, western part NE¼ sec. 35, T. 36 N., R. 6 W. (Schell Mountain quadrangle). Formation.
2. Location same as No. 1. Composite sample 11 feet long from sheared greenstone,
3. Roadcut, SE¼SE¼ sec. 35, T. 36 N., R. 6 W. (Schell Mountain quadrangle). Random near mica porphyry dike at contact between Bragdon Formation and Copley

TABLE 2. —*Analyses of samples,*

[Analyses, except gold, are semiquantitative spectrographic by E. E. Martinez and K. C. following symbols: N, not detected: L, detected but below limit of determination. percent. Gold analyses by atomic absorption are by W. L. Campbell, R. F. Hansen, reported in parts per million]

Sample No.	Laboratory No.	Field No.	Ag	As	Au	B	Ba	Be	Bi	Cd	Co	Cr	Cu	La
1	ADH—102	M—1—66	N	N	N	70	1,000	1	N	N	N	150	20	N
2	103	2	1	N	1.3	50	700	1	N	N	N	200	20	50
3	104	3	N	N	N	30	500	1	N	N	5	100	70	20
4	105	4	1	N	N	N	1,000	1	N	N	N	70	70	20
5	106	5	N	N	N	50	1,000	1	N	N	N	100	200	30
6	107	6	N	N	N	70	1,000	1	N	N	5	50	200	N
7	108	7	N	N	2.1	10	500	1	100	N	N	70	1,000	N
Limits of determination			0.5	200	.1	10	5	1	10	20	5	10	2	20
8	AEA—608	M—1—67	L	L	N	30	500	1	N	N	10	150	50	N
9	609	2	N	L	N	L	70	L	N	N	5	10	50	N
10	610	3	L	L	.1	15	300	2	N	N	10	30	20	N
11	611	4	L	N	N	50	500	1	N	N	30	100	50	N
12	612	5	N	N	N	10	300	L	N	N	10	L	L	50
13	613	6	N	N	N	50	300	1	N	N	15	100	30	N
14	614	7	L	N	N	30	500	1.5	N	N	20	150	30	N
15	615	8	N	N	N	50	300	1	N	N	20	100	30	N
16	616	9	N	N	N	100	700	1	N	N	10	70	30	N
17	617	10	L	N	.02	50	500	L	N	N	10	70	50	N
18	618	11	N	L	.02	30	700	L	N	N	15	200	500	N
19	621	15	.7	200	.06	30	300	L	N	N	5	70	20	N
20	622	16	6	300	.04	15	150	L	N	N	20	200	30	N
21	641	SM—4—67	L	N	.02	50	1,000	1	N	N	5	10	20	N
22	642	5	N	N	N	30	500	L	N	N	7	L	10	N
23	643	6	N	N	N	20	300	L	N	N	5	L	7	N
24	644	7	N	N	N	20	500	L	N	N	7	L	20	N
25	645	8	N	L	N	30	700	L	N	N	5	L	7	N
26	646	9	N	N	N	50	1,000	1	N	N	L	5	7	N
27	647	10	L	N	N	20	300	L	N	N	L	5	15	N
28	648	11	N	N	.02	30	300	L	N	N	5	5	20	N
29	649	12	N	N	.1	100	500	1	N	N	15	30	30	N
Limits of determination			.5	200	.02	10	20	1	10	20	5	5	5	20

Locations 1 through 20 shown in figure 5:

1. Roadcut in Bragdon Formation. SW¼ sec. 18, T. 34 N., R. 8 W. at junction of paved Composite of chips collected over 5-foot interval of wet, iron-stained fault
2. Approximately 100 yards northeast of No. 1. Random sample of soft altered slate.
3. Northeast of location 2 near end of spur road at Buckeye Arm of Clair Engle 4 feet.
4. Approximately 200 yards west of location 1. Composite of grab sample of fresh
5. Roadcut in Bragdon Formation west of Little Anna mine, Buckeye Creek, sec. 13, 20 feet.
6. Location same as No. 5. Sample from a 40-foot interval in soft buff altered slate.
7. Location same as No. 5. Sample from 2-foot width of soft yellow altered rock

Dog Creek (Delta) district, California

reported in the series 0.1, 0.15, 0.3, 0.5, 0.7, 1.0, 1.5, and so on, or by the following given in parts per million except for Fe, Mg, and Ca, which are given in percent. Gold T. A. Roemer and are reported in parts per million]

Mo	Mn	Nb	Ni	Pb	Sb	Sc	Sn.	Sr	V	W	Y	Zn	Zr	Fe	Mg	Ca	Ti
N	200	10	100	15	N	20	N	N	300	N	30	L	200	3	1	0.05	0.5
N	200	L	10	10	N	20	N	N	200	N	20	N	70	3	.2	L	.3
L	1,000	10	100	15	N	30	N	L	300	N	50	300	300	3	1	L	.3
5	10	10	5	10	100	5	10	100	10	50	10	200	10	.05	.05	.05	.002

Sample of sheared greenstone about 4 feet from contact with slate of the Bragdon

beginning 4 feet from contact with slate of the Bragdon Formation.
sample from iron oxide stained seams and shear zones in slate of the Bragdon Formation
Greenstone.

Minersville district, California

Watts, and are reported in the series 0.1, 0.15, 0.3, 0.5, 0.7, 1.0, 1.5, and so on, or by the Results are given in parts per million except for Fe, Mg, and Ca, which are given in E. E. Martinez, F. Michaels, R. L. Miller, M. S. Rickard, and T. A. Roemer and are

Mo	Mn	Nb	Ni	Pb	Sb	Sc	Sn	Sr	V	W	Y	Zn	Zr	Fe	Mg	Ca	Ti
N	300	--	50	20	N	50	N	100	150	N	20	N	150	7	1.5	1	0.5
N	100	--	20	500	N	50	N	N	200	N	30	N	150	7	.5	<.05	.5
N	300	--	50	20	N	50	N	N	150	N	15	N	100	10	.7	.1	.3
5	500	--	30	20	N	30	N	200	500	N	30	N	150	2	2	2	.3
15	50	--	30	20	N	30	N	N	700	N	70	N	100	3	.5	.05	.2
10	100	--	50	15	N	30	N	N	200	N	30	N	200	5	.5	.05	.5
15	500	--	30	10	N	20	N	N	500	500	10	N	100	15	.5	.15	.3
2	10	--	2	10	100	5	10	50	10	50	5	200	10	.05	.01	.01	.1
N	70	10	150	10	N	20	N	N	150	N	20	L	200	5	.5	L	.3
N	50	L	7	N	N	5	N	N	30	N	L	L	30	1.5	.1	L	.07
L	100	10	50	15	N	10	N	N	50	N	10	N	100	2	.2	L	.1
L	200	10	70	15	N	20	N	N	300	N	20	L	200	3	1	L	.5
N	500	L	7	15	N	5	N	N	10	N	15	L	70	1.5	.15	L	.07
N	700	10	100	20	N	10	N	N	100	N	15	N	100	3	1	.1	.3
L	500	10	70	15	N	15	N	N	200	N	20	200	150	3	1	.1	.5
L	700	10	70	15	N	20	N	100	200	N	20	L	200	3	1.5	.5	.5
L	100	10	70	15	N	20	N	N	200	N	30	L	200	3	.7	L	.5
L	300	10	70	15	N	15	N	N	200	N	20	L	150	3	.5	L	.3
N	200	L	100	L	N	30	N	N	150	N	20	L	50	5	.7	L	.3
N	100	10	7	100	N	7	N	N	200	N	10	L	100	3	.3	L	.3
N	1,000	L	150	10	N	30	N	L	300	N	15	L	30	3	2	.07	.3
N	30	L	7	10	N	10	N	L	30	N	15	L	150	5	.2	L	.15
N	100	L	7	L	N	20	N	L	50	N	50	N	150	3	.3	.07	.2
N	70	L	5	L	N	30	N	L	70	N	30	N	150	2	.2	.07	.2
N	100	L	7	L	N	10	N	L	30	N	30	N	100	2	.3	.1	.2
15	30	L	7	L	N	7	N	100	20	N	15	N	100	5	.05	.07	.15
L	30	10	5	L	N	10	N	L	70	N	15	N	150	3	.5	L	.2
N	30	L	7	L	N	7	N	N	30	N	15	N	100	1.5	.15	.05	.15
N	300	L	30	15	N	10	N	L	30	N	30	L	100	2	.5	.1	.15
N	700	L	50	10	N	20	N	100	100	N	20	L	200	2	.2	.1	.3
5	10	10	5	10	100	5	10	100	10	50	10	200	10	.05	.05	.05	.002

road in Buckeye Creek with spur to Buckeye Arm of Clair Engle (Trinity) Lake.
zone.
at contact with weathered dike rock, exposed in portal of prospect adit.
(Trinity) Lake. Sample cut from gouge zone in slate over a distance of approximately

unaltered slate of the Bragdon Formation.
T. 34 N., R. 9 W. Composite of chips of hard slate collected over a distance of about

exposed in small pit beside road.

TABLE 2.—*Analyses of samples,*

Locations 1 through 20 shown in figure 5—Continued

8–13. Samples 8 through 13 collected from cuts along logging road on east side
   8. Composite sample taken over a distance of 1 foot each side of contact of
   9. Composite sample of quartz vein 10 feet wide.
   10. Sample from soft, weathered dike cutting Bragdon Formation. Sample 6
   11. Location same as No. 10. Sample of slate from Bragdon Formation over
   12. Random sample of weathered dike over width of 200 feet.
   13. Near junction of log road with main paved road. Weathered dike intruding
      contact.
   14. Approximately 125 feet east of Little Elsie mine, south-central part sec. 18,
      sheared iron-stained slate of Bragdon Formation.
   15. Sample about 160 feet east of Little Elsie. Brecciated slate.
16–18. Samples 16 through 18 from contact zone between Bragdon and Copley
      mine, Whitney Gulch, SE¼ sec. 18, T. 34 N., R., R. 8 W.
   16. Brecciated slate west of contact. Composite of chips collected over a dis-
   17. Brecciated slate west of contact. Composite of chips collected over
   18. Composite sample of partly weathered pyritic greenstone collected over a
   19. Roadcut, SW¼NW¼ sec. 16, T. 34 N., R. 8 W., exposing contact between
      some dacite porphyry. Sample is composite of chips from a small
   20. Composite taken over a 10-foot-wide zone of fine-grained iron-stained rock
21–27. Samples 21 through 27 are from a quartz porphyry intrusion into slate of
      miles southeast of Five Pines mine and 0.3 mile northwest of Halfway
      Distances are measured northwest along road starting at centerline of
   21. Distance, 128 feet. Sample of soft iron oxide stained seam, 2 inches wide.
   22. Distance, 7 to 23 feet. Composite of chips from fine-grained, bleached, brec-
   23. Distance, 23 to 65 feet. Composite of chips from hard, very fine grained quartz
   24. Distance, 90 to 120 feet. Composite of chips from slightly iron stained quartz
   25. Distance, 128 to 138 feet. Composite of chips from gray brecciated pyriti-
   26. Distance, 270 feet. Sample of shear zone in brecciated fine-grained quartz
   27. Distance, 280 feet. Sample same as No. 26.
28, 29. Samples 28 and 29 are from small outcrop of quartz porphyry exposed in
      4015 on county-line ridge, NW¼ sec. 34, T. 35 N., R. 7 W., Schell Moun-
   28. Composite of chips from quartz porphyry.
   29. Sample of soft sheared weathered quartz prophyry.

TABLE 3.—*Analyses of samples from*

[Analyses, except gold, are semiquantitative spectrographic by K. C. Watts, and are
symbols: N, not detected; L, detected but below limit of determination. Results are
analyses by atomic absorption are by W. L. Campbell, R. L. Miller, M. S. Rickard, and

Sample No.	Laboratory No.	Field No.	Ag	As	Au	B	Ba	Be	Bi	Cd	Co	Cr	Cu	La
1	AEA—690	M—22—67	15	L	0.4	L	N	L	L	N	7	15	20	N
2	691	23	2	200	2	L	N	L	L	N	10	15	20	N
3	692	24	50	200	46	L	N	L	30	N	15	30	20	N
4	689	H—1—67	10	N	46	L	30	L	L	N	L	10	30	N
Limits of determination			.5	200	.02	10	20	1	10	20	5	5	20	5

1–3. Samples 1 through 3 are from an opencut at the Globe mine. Cut is on west
   Minersville quadrangle. Cut exposes three quartz veins that strike N. 75° E.,
   1. Lower vein, 2 feet wide. Composite sample cut perpendicular to dip.
   2. Middle vein, 6 feet wide. Composite sample cut perpendicular to dip.
   3. Upper vein. Composite sample cut perpendicular to dip.
   4. Carl No. 1 prospect. Shallow pit 1,000 feet south of Little East Fork of Canyon
      Sample cut perpendicular to dip of shattered quartz vein 18 inches thick in

TABLE 4.—*Analyses of samples,*

[Analyses, except gold, are semiquantitative spectrographic by E. E. Martinez and K. C.
the following symbols: N, not detected; L, detected but below limit of determination.
percent. Gold analyses by atomic absorption are by W. L. Campbell, R. F. Hansen,
reported in parts per million]

Sample No.	Laboratory No.	Field No.	Ag	As	Au	B	Ba	Be	Bi	Cd	Co	Cr	Cu	La
1	ADH—081	SB—7—66	N	2,000	0.3	100	2,000	N	N	N	20	300	200	30
2	082	8	N	1,000	N	30	700	N	N	N	150	200	300	50
3	083	9	N	N	N	100	3,000	N	N	N	20	70	70	20
4	084	10	N	N	N	70	2,000	N	N	N	N	2,000	200	30
5	085	11	N	N	N	70	2,000	N	N	N	5	1,500	150	50
Limits of determination			.5	200	.1	10	5	1	10	20	5	10	2	20

Minersville district, California—Continued

Buckeye Creek central part sec. 18, T. 34 N., R. 8 W.
altered andesite(?) dike with slate of the Bragdon Formation.

inches wide perpendicular to contact.
distance of 1 foot perpendicular to contact.

Bragdon Formation. Random sample of dike and slate of Bragdon Formation at

T. 34 N., R. 8 W. Fault zone about 3 feet wide. Random sample of brecciated and

Formations exposed in roadcuts approximately a quarter of a mile east of Little Elsie

tance of 35 feet, beginning 70 feet west of contact.
a distance of 35 feet west from the contact.
distance of 8 feet east from the contact.
Copley Greenstone and slate of the Bragdon Formation. Contact is a sheared zone with
siliceous (vein quartz?) block in slate.
in shear zone. Not certain whether greenstone or siliceous slate.
the Bragdon Formation exposed in roadcuts in Van Ness Creek approximately 1.4
Gulch, undivided SW¼, T. 35 N., R. 7 W. (Schell Mountain quadrangle, California).
second creek northwest of Halfway Gulch.

ciated quartz porphyry.
porphyry containing very fine grained pyrite.
porphyry.
ferous iron oxide stained quartz porphyry.
porphyry.

roadcut in head of Feany Gulch, approximately 3,500 feet north-northwest of B.M.
tain quadrangle.

near the Globe mine, California

reported in the series 0.1, 0.15, 0.3, 0.5, 0.7, 1.0, 1.5, and so on, or by the following
given in parts per million except for Fe, Mg, and Ca, which are given in percent. Gold
T. A. Roemer and are reported in parts per million]

Mo	Mn	Nb	Ni	Pb	Sb	Sc	Sn	Sr	V	W	Y	Zn	Zr	Fe	Mg	Ca	Ti
N	200	L	5	20	N	7	N	N	50	N	10	N	70	1.5	0.7	0.1	0.15
N	300	L	7	70	N	10	N	N	100	N	10	N	30	1	.5	.07	.2
N	200	10	7	500	N	10	N	N	200	20	15	N	50	3	.7	.1	.3
N	50	L	5	100	N	5	N	N	50	N	N	N	N	.03	.3	.05	.07
5	10	10	5	10	100	5	10	100	10	50	10	200	10	.05	.05	.05	.002

side of ridge at altitude of approximately 6,400 feet, T. 35 N., R. 10 W. (undivided)
dip 45° S., in sheared hornblende schist.

Creek, altitude 4,400 feet, T. 35 N., R. 10 W. (undivided), Helena quadrangle.
amphibolite. Strike of vein N. 70 E., dip 45° S.

Liberty district, California

Watts, and are reported in the series 0.1, 0.15, 0.3, 0.5, 0.7, 1.0, 1.5, and so on, or by
Results are given in parts per million except for Fe, Mg, and Ca, which are given in
E. E. Martinez, F. Michaels, R. L. Miller, M. S. Rickard, and T. A. Roemer, and are

Mo	Mn	Nb	Ni	Pb	Sb	Sc	Sn	Sr	V	W	Y	Zn	Zr	Fe	Mg	Ca	Ti
10	500	--	70	20	N	50	N	N	150	N	30	N	--	10	0.5	0.2	0.5
20	2,000	--	500	20	N	50	N	N	150	N	50	200	--	15	1	.2	.7
N	500	--	30	20	N	30	N	N	100	N	N	N	--	10	.5	.5	.5
15	500	--	300	20	N	20	N	N	100	N	20	200	--	10	7	.2	.7
30	2,000	--	200	15	N	50	N	N	200	N	30	N	--	10	5	.7	1.0
2	10	--	2	10	100	5	10	50	10	50	5	200	10	.05	.01	.01	.001

TABLE 4.—*Analyses of samples,*

Sample No.	Labora- tory No.	Field No.	Ag	As	Au	B	Ba	Be	Bi	Cd	Co	Cr	Cu	La
6	AEA—626	SB—3—67	N	N	N	30	500	1	N	N	10	100	30	N
7	627	4	N	N	N	50	500	1.5	N	N	20	150	50	N
8	629	6	N	N	N	100	700	1	N	N	15	150	50	N
9	630	7	N	N	N	50	500	3	N	N	15	100	50	N
10	631	8	L	N	N	100	1,000	3	N	N	15	70	50	N
11	632	9	N	N	N	100	700	1.5	N	N	10	50	50	N
12	633	10	L	N	N	150	1,000	3	N	N	20	100	100	N
13	634	11	N	N	.2	100	500	1	N	N	15	150	30	N
Limits of determination.......			0.5	200	.02	10	20	1	10	20	5	5	5	20

1—5. Samples 1 through 5 from cuts along Eddy Gulch road to Black Bear California.
  1. First gully south of West Fork Eddy Gulch. Fault zone cut by andesitic dikes
  2. First gully south of West Fork Eddy Gulch. Sample from gently dipping shear
  3. Second gully south of West Fork Eddy Gulch, or first gully west of Humpback rock. Four-foot sample along dip.
  4. 3,200 feet S. 68° E. from Black Bear summit. Shear zone in siliceous schist.
  5. 2,300 feet S.54° E. from Black Bear summit. Sample 2 feet long in a 50-foot-wide
6—13. Samples 6 through 13 from roadside exposure and cuts in graphitic mica discontinuous veinlets of quartz. Stained with iron oxide in places, and
  6. Exposure in gully east of roadcut. Combined random chips of sheared phyllite.
7—13. Samples 7 through 13 each 25 feet long, perpendicular to dip, beginning 65 feet
  7. 0—25 feet.
  8. 25—50 feet.
  9. 50—75 feet.
  10. 75—100 feet.
  11. 100—125 feet.
  12. 125—150 feet.
  13. 150—175 feet.

TABLE 5. —*Analyses of samples,*

[Analyses, except gold, are semiquantitative spectrographic by E. E. Martinez, and are symbols: N, not detected; L, detected but below limit of determination. Results are analyses by atomic absorption are by W. L. Campbell, R. L. Miller, M. S. Rickard, and

Sample No.	Labora- tory No.	Field No.	Ag	As	Au	B	Ba	Be	Bi	Cd	Co	Cr	Cu	La
1	AEA—681	WI—5—67	0.7	N	N	10	500	1.5	N	N	10	70	150	N
2	682	6	.5	N	0.04	L	200	L	N	N	7	30	50	N
3	683	7	3	N	N	15	500	1	N	N	5	50	150	50
Limits of determination...........			.5	200	.02	10	20	1	10	20	5	5	5	20

All samples from cuts along Evans Creek Road, south edge NW¼ sec. 34, T. 34 S.,

1. Composite chip sample of 1-inch-wide black gougy seam containing fine-grained
2. Composite chip sample taken perpendicular to dip of fracture zone over a distance of
3. Composite of chips from minor narrow (1-inch) quartz-mica pyrite(?) veinlets that

Liberty district, California—Continued

Mo	Mn	Nb	Ni	Pb	Sb	Sc	Sn	Sr	V	W	Y	Zn	Zr	Fe	Mg	Ca	Ti
5	500	10	70	10	N	10	N	N	100	N	20	L	100	3	1	0.1	0.3
10	700	15	150	10	N	15	N	N	200	N	20	L	150	3	1.5	.2	.3
10	500	10	150	15	N	10	N	N	100	N	15	L	150	2	1.5	.07	.3
7	700	15	100	15	N	20	N	N	150	N	20	L	150	2	1.5	.1	.5
15	300	20	70	15	N	30	N	N	200	N	20	N	200	3	1	.1	.5
N	500	15	50	10	N	10	N	N	70	N	20	200	70	10	.5	.05	.2
10	500	15	150	15	N	20	N	N	200	N	30	L	150	3	1.5	.1	.3
7	500	15	150	10	N	20	N	L	150	N	20	200	150	7	1.5	.1	.3
5	10	10	5	10	100	5	10	100	10	50	10	200	10	.05	.05	.05	.002

summit in undivided northwestern part of T. 39 N., R. 11 W., Sawyers Bar quadrangle,

on north side of gully. Combined sample from three shear planes containing some quartz.
zone in schist few feet south of gully. White clay iron oxide staining and quartz.
Gulch, in bank south of gully. Iron oxide stained shear zone (N. 80 W., 45° S.) in dike

fault zone. Black gouge with quartz, iron oxide staining.
schist and phyllite; strikes N. 10° W., dips 60° W. to vertical. Thin quartz laminae and
contains pyrite. East edge SE¼ sec. 5, T. 39 N., R. 10 W., Sawyers Bar quadrangle.

N. 75° W. from gully of sample 6. Combined chips.

Evans Creek, Oreg.

reported in the series 0.1, 0.15, 0.3, 0.5, 0.7, 1.0, 1.5, and so on, or by the following
given in parts per million except for Fe, Mg, and Ca, which are given in percent. Gold
T. A. Roemer and are reported in parts per million]

Mo	Mn	Nb	Ni	Pb	Sb	Sc	Sn	Sr	V	W	Y	Zn	Zr	Fe	Mg	Ca	Ti
20	300	10	10	L	N	15	N	500	300	N	20	L	150	3	0.7	2	0.3
15	300	10	20	15	N	15	N	300	300	N	15	L	100	3	.3	2	.3
5	200	10	10	L	N	15	N	100	200	N	30	L	50	3	.2	.2	.3
5	10	10	5	10	100	5	10	100	10	50	10	200	10	.05	.05	.05	.002

R. 3 W. (Wimer quadrangle):

pyrite. Seam occurs at hanging-wall contact of fracture zone. Strike N. 80° W., dip 20° SE.
about 4 feet.
strike N. 45° E., dip 40° NW.

TABLE 6. —*Analyses of samples,*

[Analyses, except gold, are semiquantitative spectrographic by E. E. Martinez, and are symbols: N, not detected; L, detected but below limit of determination. Results are analyses by atomic absorption are by W. L. Campbell, R. L. Miller, M. S. Rickard, and

Sample No.	Laboratory No.	Field No.	Ag	As	Au	B	Ba	Be	Bi	Cd	Co	Cr	Cu	La
1	AEA–671	GL–11–67	L	N	0.2	L	50	L	N	N	20	150	50	N
2	672	12A	N	L	.8	L	30	N	N	N	20	100	50	N
3	673	12B	N	N	.8	N	20	L	N	N	15	100	50	N
4	674	12C	N	N	N	10	20	L	N	N	15	20	30	N
Limits of determination			0.5	200	.02	10	20	1	10	20	5	5	5	20

All samples from Greenback mine, SW¼SW¼ sec. 33, T. 33 S., R. 5 W., Glendale, Oreg., quadrangle:

1. Grab sample across toe of dump at mouth of small opencut above and east of old tailings piles.
2. Channel sample, 10 feet long, of weathered greenstone from north wall of cut.

TABLE 7. —*Analyses of samples,*

[Analyses, except gold, are semiquantitative spectrographic by E. E. Martinez and are symbols: N, not detected; L, detected but below limit of determination; G, greater are given in percent. Gold analyses by atomic absorption are by W. L. Campbell, R. L.

Sample No.	Laboratory No.	Field No.	Ag	As	Au	B	Ba	Be	Bi	Cd	Co	Cr	Cu	La
1	AEA–655	GAL–2–67	5	N	0.2	10	G(5,000)	L	L	N	10	5	300	N
2	656	3	10	L	.1	L	G(5,000)	N	N	N	7	L	100	N
3	657	4	10	1,500	.4	15	G(5,000)	N	N	N	5	L	150	N
4	658	5	30	700	.2	10	G(5,000)	N	N	N	5	L	100	N
5	659	6	70	700	2.4	10	G(5,000)	L	15	N	N	7	150	N
6	660	7	L	N	.08	10	5,000	L	L	N	N	5	30	N
7	662	18	L	N	.04	15	500	L	N	N	50	50	N
8	663	19	.5	N	.02	L	200	L	N	N	10	20	30	N
9	664	20	L	N	N	30	500	L	N	N	15	L	30	N
Limits of determination			.5	200	.02	10	20	1	10	20	5	5	5	20

Locations 1 through 9 shown in figure 6:

1. Combined sample of chips taken at 5-foot intervals over a distance of approximately river level on north side of Rogue River. Sample taken westward from dioritic
2–6. Samples 2 through 6 are combined chip samples cut from bleached and iron with dioritic intrusive body exposed at portal of upper adit, near shaft,
2. Hard silicified barite rock, 25 feet wide (beginning at portal).
3. Soft brown porous zone, 5 feet wide.
4. Hard, gray, silicified brecciated dacite porphyry(?) containing very fine grained
5. Barite rock, little apparent pyrite. In part brown and porous, 8 feet wide.
6. Mostly white, altered metavolcanic rock, with some hard gray bodies, and a little
7, 8. Samples 7 and 8 collected from roadcut northeast side Rocky Gulch. Galice-wide of brown-stained sheared greenstone bleached in places owing to blocks of very fine grained siliceous rock containing very fine grained pyrite.
7. Combined chips sample 55 feet long of altered greenstone.
8. Combined chips from siliceous blocks.
9. Sailor Jack Creek, SW¼NW¼ sec, 10, T. 35 S., R. 8 W. Sample of 10-foot

Greenback district, Oregon

reported in the series 0.1, 0.15, 0.3, 0.5, 0.7, 1.0, 1.5, and so on, or by the following given in parts per million except for Fe, Mg, and Ca, which are given in percent. Gold T. A. Roemer and are reported in parts per million]

Mo	Mn	Nb	Ni	Pb	Sb	Sc	Sn	Sr	V	W	Y	Zn	Zr	Fe	Mg	Ca	Ti
N	1,000	L	100	L	N	30	N	200	150	N	20	N	70	3	3	3	0.3
N	1,000	L	70	L	N	20	N	200	150	N	20	N	100	3	1.5	1	.3
N	700	L	50	N	N	20	N	100	100	N	20	N	150	3	1.5	.7	.3
N	700	10	7	L	N	30	N	L	100	N	50	N	300	3.	.7	.2	.3
5	10	10	5	10	100	5	10	100	10	50	10	200	10	.05	.05	.05	.002

3. Channel sample, 12 feet long, of weathered greenstone from east face of cut.
4. Channel sample, 12 feet long, of weathered greenstone from south wall of cut.

Galice district, Oregon

reported in the series 0.1, 0.15, 0.3, 0.5, 0.7, 1.0, 1.5, and so on, or by the following than value shown. Results are given in parts per million except for Fe, Mg, and Ca, which Miller, M. S. Rickard, and T. A. Roemer and are reported in parts per million]

Mo	Mn	Nb	Ni	Pb	Sb	Sc	Sn	Sr	V	W	Y	Zn	Zr	Fe	Mg	Ca	Ti
L	300	L	7	100	N	10	N	500	100	N	10	L	70	2	0.5	0.5	0.3
7	15	L	5	30	L	7	N	700	30	N	L	L	50	3	L	L	.2
30	15	L	L	50	150	7	N	500	70	N	L	L	20	5	.02	L	.1
30	20	L	5	100	100	5	L	1,500	50	N	L	700	50	5	.02	L	.15
15	10	L	L	700	L	5	30	2,000	150	N	L	L	50	3	.02	L	.15
L	20	L	L	150	N	15	N	200	150	N	10	N	70	3	.07	L	.3
5	50	L	5	200	N	20	N	N	200	N	L	N	70	5	.7	L	.3
L	30	L	7	20	N	10	N	N	70	N	L	N	70	3	.07	L	.3
L	300	L	7	10	N	20	N	200	200	N	15	N	70	3	.7	1	.3
5	10	10	5	10	100	5	10	100	10	50	10	200	10	.05	.05	.05	.002

mately 250 feet perpendicular to strike of a zone of pyritized greenstone exposed at dike at slate-greenstone contact. Almeda mine, SE¼SE¼ sec. 13, T. 34 S., R. 8 W. oxide stained greenstone exposed in roadcuts above river level, starting at west contact Almeda mine, SE¼SE¼ sec. 13, T. 34 S., R. 8 W. Samples collected from east to west.

pyrite, 75 feet wide.

brown oxidized barite rock, 160 feet wide.
Rogue Formation contact. SE¼SE¼ sec. 26, T. 34 S., R. 8 W. Zone 50 to 75 feet weathering of pyrite. Zone at a few places contains hard, manganese oxide stained

zone altered bleached iron oxide stained metavolcanic rock at slate-greenstone contact.

TABLE 8.—*Principal lode gold mines of the Klamath Mountains province, California and Oregon*

Number on pl. 2	Mine	District	County	Approximate production (dollars)
		MINES IN OREGON		
1	Chieftain and Continental	Myrtle Creek	Douglas	>200,000
2	Benton	Galice	Josephine	500,000-800,000
3	Gold Bug	do	do	750,000
4	J.C.L.	do	do	100,000
5	Almeda	do	do	31,000
6	Robertson (Bunker Hill)	do	do	138,000
7	Silent Friend	Greenback	do	34,000
8	Dorothea	do	do	50,000
9	Greenback	do	do	3,500,000
10	John Hall group	do	do	90,000
11	Daisy (Hammersley)	do	do	250,000
12	Rainbow (Siskron)	Takilma	do	46,500
13	Robert E	Illinois River	Curry	>100,000
14	Sylvanite	Rogue River—Applegate	Jackson	700,000
15	Lucky Bart	do	do	200,000
16	Gold Hill pocket	do	do	700,000
17	Braden	do	do	30,000
18	Revenue pocket	do	do	100,000
19	Town	do	do	90,000
20	Opp	do	do	>100,000
21	Oregon Belle	do	do	250,000
22	Ashland	Ashland	do	1,300,000
23	Shorty Hope	do	do	30,000
24	Steamboat	Upper Applegate	do	>350,000
		MINES IN CALIFORNIA		
25	Jillson	Yreka—Fort Jones	Siskiyou	800,000
26	Hicks (China Gulch)	do	do	>30,000
27	Indian Girl	do	do	300,000
28	Commodore	do	do	230,000
29	Scott Bar	do	do	>41,000
30	Eliza	do	do	750,000
31	Schroeder	do	do	>100,000
32	Mono	do	do	1,000,000
33	Katie May	do	do	>70,000
34	Osgood	do	do	>40,000
35	Golden Eagle	do	do	500,000-1,000,000
36	Franklin	do	do	90,000
37	Morrison and Carlock	do	do	500,000
38	Blind Lode	do	do	40,000
39	Independence	Happy Camp	do	300,000
40	Siskon	do	do	>3,600,000
41	Dewey	Callahan	do	900,000
42	Hathaway	do	do	>60,000
43	Cummings (McKeen)	do	do	>500,000
44	Highland	Liberty	do	350,000
45	Advance	do	do	>250,000
46	Lanky Bob	do	do	50,000
47	Cleaver (Brown Bear)	do	do	30,000
48	Hickey	do	do	35,000-60,000
49	Uncle Sam	do	do	75,000
50	Ball (California Consolidated)	do	do	473,500
51	Mountain Laurel	do	do	500,000
52	Evening Star (Union)	do	do	79,000-90,000
53	Klamath	do	do	>600,000
54	Humpback (Fagundez)	do	do	260,000
55	Black Bear	do	do	>3,000,000
56	Hansen	Cecilville	do	20,000
57	Gilta	do	do	>500,000
58	Knownothing	do	do	>100,000
59	King Solomon	do	do	679,000

TABLE 8.—*Principal lode gold mines of the Klamath Mountains province, California and Oregon*—Continued

Number on pl. 2	Mine	District	County	Approximate production (dollars)
		MINES IN CALIFORNIA—Continued		
60	Sherwood	New River—Denny	Trinity	100,000
61	Uncle Sam	— do	— do	60,000
62	Ridgeway	— do	— do	80,000
63	Hard Times	— do	— do	20,000
64	Excelsior	— do	— do	160,000
65	Mountain Boomer	— do	— do	350,000
66	Dorleska	Trinity Center	— do	200,000
67	Golden Jubilee	— do	— do	250,000
68	Chapman	— do	— do	20,000
69	Blue Jay	— do	— do	>60,000
70	Nonpareil	— do	— do	20,000
71	Headlight	— do	— do	500,000
72	Bonanza King	— do	— do	1,250,000
73	Golden Chest	Canyon Creek—East Fork	— do	200,000
74	Alaska	— do	— do	600,000
75	Globe	— do	— do	>700,000
76	Ralston	— do	— do	40,000
77	Silver Grey	— do	— do	40,000– 45,000
78	Enterprise	— do	— do	>350,000
79	North Star	— do	— do	200,000
80	Fountainhead	— do	— do	>50,000
81	Ozark	— do	— do	60,000
82	Five Pines	Minersville	— do	275,000
83	Fairview	— do	— do	500,000
84	Venecia	French Gulch—Deadwood	— do	500,000
85	Last Chance	— do	— do	20,000
86	Gifford	— do	— do	75,000
87	Brown Bear	— do	— do	8,000,000
88	Amy Balch	— do	— do	60,000
89	Niagara	— do	Shasta	>1,000,000
90	Summit and Montezuma	— do	— do	200,000
91	Brunswick	— do	— do	100,000
92	Washington	— do	— do	>2,000,000
93	Milkmaid and Franklin	— do	— do	350,000
94	Highland	— do	— do	400,000
95	American	— do	— do	300,000
96	Gladstone	— do	— do	>3,000,000
97	Uncle Sam	Backbone	— do	>1,000,000
98	Mad Mule	Whiskeytown	— do	1,000,000
99	Truscott	— do	— do	60,000
100	Eldorado	— do	— do	25,000
101	Gambrinus	Shasta—Redding	— do	>127,000
102	Mount Shasta	— do	— do	>178,000
103	National (Veteran, Forbes)	Old Diggings	— do	200,000
104	Texas Consolidated	— do	— do	750,000
105	Central	— do	— do	500,000
106	Reid	— do	— do	2,500,000
107	Midas	Harrison Gulch	— do	>4,000,000
108	Kelly	Hayfork	Trinity	>100,000

INDEX

[Italic page numbers indicate major references]

☆ U.S. GOVERNMENT PRINTING OFFICE: 1970 O—388-399

www.ingramcontent.com/pod-product-compliance
Lightning Source LLC
Chambersburg PA
CBHW031951190326
41519CB00007B/759